# The Business of Plant Breeding

## Market-led Approaches to New Variety Design in Africa

The e-learning material is available at
http://www.cabi.org/openresources/93814 and also
on a USB stick that is included with this volume.

# The Business of Plant Breeding

## Market-led Approaches to New Variety Design in Africa

*Edited by*

**G.J. Persley**

*Global Change Institute, University of Queensland, St Lucia, Australia*
*The Doyle Foundation, Glasgow, Scotland, UK*

and

**V.M. Anthony**

*Syngenta Foundation for Sustainable Agriculture, Basel, Switzerland*

**CABI is a trading name of CAB International**

CABI
Nosworthy Way
Wallingford
Oxfordshire OX10 8DE
UK

Tel: +44 (0)1491 832111
Fax: +44 (0)1491 833508
E-mail: info@cabi.org
Website: www.cabi.org

CABI
745 Atlantic Avenue
8th Floor
Boston, MA 02111
USA

Tel: +1 (617)682-9015
E-mail: cabi-nao@cabi.org



A catalogue record for this book is available from the British Library, London, UK.

**Library of Congress Cataloging-in-Publication Data**

Names: Persley, G. J., editor. | Anthony, V. M. (Vivienne M.), editor.
Title: The business of plant breeding : market-led approaches to new variety design in Africa / edited by G.J. Persley, V.M. Anthony.
Description: Boston, MA : CABI, [2017] | Includes bibliographical references and index.
Identifiers: LCCN 2017037317 (print) | LCCN 2017047037 (ebook) | ISBN 9781786393838 (open access) | ISBN 9781786393814 (hardback : alk. paper)
Subjects: LCSH: Plant breeding--Africa. | Plant varieties--Africa.
Classification: LCC SB123.25 (ebook) | LCC SB123.25 B87 2017 (print) | DDC 631.5/3096--dc23
LC record available at https://lccn.loc.gov/2017037317

ISBN-13: 9781786393814

Commissioning editor: Dave Hemming
Editorial assistant: Alexandra Lainsbury
Production editor: James Bishop

Typeset by SPi, Pondicherry, India
Printed and bound in the UK by Bell and Bain Ltd, Glasgow

# Contents

# Contributors and Editors

**Vivienne M. Anthony**, Syngenta Foundation for Sustainable Agriculture, WRO-1002.11.54, PO Box 4002, Basel, Switzerland. E-mail: Vivienne.Anthony@syngenta.com

**Rowland Chirwa**, CIAT (International Center for Tropical Agriculture), Chitedze Agricultural Research Station (CARS), PO Box 158, Lilongwe, Malawi. E-mail: r.chirwa@cgiar.org

**Agyemang Danquah**, West Africa Centre for Crop Improvement (WACCI), College of Basic and Applied Sciences, University of Ghana, Legon, PMB LG 30, Ghana. E-mail: adanquah@wacci.edu.gh

**Eric Y. Danquah**, West Africa Centre for Crop Improvement (WACCI), College of Basic and Applied Sciences, University of Ghana, Legon, PMB LG 30, Ghana. E-mail: edanquah@wacci.edu.gh

**Appolinaire Djikeng**, Centre for Tropical Livestock Genetics and Health (CTLGH) The Roslin Institute, The University of Edinburgh, Edinburgh, Scotland, U.K. E-mail: appolinaire.djikeng@ctlgh.org

**Shimelis Hussein**, African Centre for Crop Improvement (ACCI), University of KwaZulu-Natal, P/Bag X01, Scottsville 3209, Pietermaritzburg, South Africa. E-mail: Shimelish@ukzn.ac.za

**Paul M. Kimani**, Plant Breeding and Biotechnology Program, Department of Plant Science and Crop Protection, College of Agriculture and Veterinary Sciences, University of Nairobi, PO Box 29053-00625, Nairobi, Kenya. E-mail: pmkimani@uonbi.ac.ke

**Heather Merk**, Syngenta Seeds Inc., 11055 Wayzata Blvd, Minnetonka, MN 55305, USA. E-mail: heather.merk@syngenta.com

**Gabrielle J. Persley**, Global Change Institute, University of Queensland, St Lucia, Queensland 4072, Australia and The Doyle Foundation, 45 St Germains, Bearsden, Glasgow, G61 2RS, Scotland, UK. E-mail: g.persley@doylefoundation.org and g.persley@cgiar.org

**Jean Claude Rubyogo**, Seed Systems and Agricultural Technology Transfer, International Center for Tropical Agriculture (CIAT), PO Box 2704, Arusha, Tanzania. E-mail: j.c.rubyogo@cgiar.org

**Ivan Rwomushana**, International Centre for Insect Physiology and Ecology (ICIPE), Nairobi, Kenya. E-mail: Irwomushana@icipe.org

**Jonathan L. Shoham**, c/o Syngenta Foundation for Sustainable Agriculture, WRO-1002.11.54, PO Box 4002, Basel, Switzerland. E-mail: jonathanl.shoham@gmail.com

**Pangirayi Tongoona**, West Africa Centre for Crop Improvement (WACCI), College of Basic and Applied Sciences, University of Ghana, Legon, PMB LG 30, Ghana. E-mail: ptongoona@wacci.edu.gh

**Nassser Yao**, Biosciences eastern and central Africa (BecA), International Livestock Research Institute (ILRI), PO Box 30709, Nairobi 00100, Kenya. E-mail: n.yao@cgiar.org

# Foreword

As African agriculture is transforming from subsistence farming to more market-led systems and small producers generate food surpluses to sell, products must meet market demand. Success in demand-led breeding depends on: the setting of breeding targets and quantitative goals; new varieties reaching and fulfilling client expectations; a development strategy designed for each new variety; a delivery investment plan being in place; and emphasis given to the views of both farmers and consumers from rural and urban areas. Success in demand-led plant breeding will be determined by the adoption and use of the new varieties that meet the market-led demands throughout crop value chains.

An Alliance for Food Security in Africa that was formed by the Australian Centre for International Agricultural Research, the Crawford Fund, Australia and the Syngenta Foundation for Sustainable Agriculture has joined with African, Australian and international research institutes and universities in contributing towards the transformation of African agriculture in the area of crop improvement. The first project supported by this Alliance is on 'Demand-led Plant Variety Design' in order to advocate more emphasis on the identification and inclusion of market-demanded characteristics in the design of new plant varieties. The participants, both in Africa and internationally, are identifying and sharing best practices in demand-led plant breeding from both private and public sector breeding programmes in a range of tropical crops. These best practices will be taught to the next generation of African plant breeders through the continent's universities in order to ensure a greater uptake of improved varieties, enhanced livelihoods for farmers and food security across Africa. The project is administered by the Global Change Institute of the University of Queensland, on behalf of the Alliance. The members of the Alliance thank the University of Queensland, particularly Professors Ove Hoegh-Gulberg, Director and Karen Hussey, Deputy Director of the Global Change Institute (GCI), Professor Bill Bellotti and Dr Grace Muriuki of the CGI Food Systems Program for their support and contributions to the Demand-led Plant Variety

Design Project. The support of Dr Andrew Bennett and of the Doyle Foundation, Scotland, in the preparation of this volume is also gratefully acknowledged. The advice of Dr Joe DeVries was particularly helpful in establishing the project.

An early outcome of the Project is this handbook – *The Business of Plant Breeding: Market-led Approaches to New Variety Design in Africa*. It is the result of the work of an educators' group of plant breeders from national, regional and international research institutes and universities throughout Africa that have responsibilities for postgraduate education and the professional development of plant breeders, supported by plant breeders with private sector breeding experience.

The members of the Alliance would like to thank the members of the educators' group, particularly the West African Centre for Crop Improvement (WACCI) of the University of Ghana who convened the Pan-Africa educators' group, and the African Centre for Crop Improvement (ACCI) of the University of KwaZulu-Natal South Africa and Biosciences eastern and central Africa (BecA) at the International Livestock Research Institute in Nairobi, Kenya, who coordinated activities in southern and eastern Africa, respectively. We further thank all of the members of the educators' group who are listed on the Contributors page and who have lent their experience to the work of the group and written the specialist chapters that are included in this book. We are grateful to Syngenta Seeds for enabling Dr Heather Merk to participate as a resource person for the educators' group and to contribute her experience in the professional development of plant breeders.

We would also like to thank Syngenta Seeds and Market Edge for kindly making available a state-of-the art interactive education tool for designing new varieties, and Syngenta Seeds for their insights and their education tool for understanding and creating new variety breeding investment cases.

We trust that this book will prove useful in both the formal postgraduate education of plant breeders and in their continuing professional development, not only in Africa, but also on other continents, where agriculture must transform to meet new challenges and to respond to emerging markets, as well as the challenges of ensuring food security, changing climate and evolving consumer preferences.

**Denis Blight**
Chief Executive
The Crawford Fund
Canberra, Australia

**Marco Ferroni**
Executive Director
Syngenta Foundation for Sustainable Agriculture
Basel, Switzerland

**Mellissa Wood**
General Manager, Global Program
Australian Centre for International Agricultural Research
Canberra, Australia

# Executive Summary

Vivienne M. Anthony[1]* and Gabrielle J. Persley[2†]

[1]Syngenta Foundation for Sustainable Agriculture, Basel, Switzerland; [2]Global Change Institute, University of Queensland, St Lucia, Australia and The Doyle Foundation, Glasgow, UK

## Introduction

Demand-led plant breeding combines the best practices in market-led new variety design with innovative plant breeding methods and integrates both of these with the best practices in business as a new way of approaching the business of plant breeding to deliver benefits.

*The Business of Plant Breeding* is the result of a study on demand-led plant variety design for changing markets in Africa, the purpose of which is to identify and share best practices in demand-led plant breeding from private and public sector breeding programmes worldwide. The intended audiences are professionals in plant breeding and related areas, such as seed production, who have interests in developing and disseminating new plant varieties as a way to increase productivity and profitability in crop agriculture, especially in Africa. The volume is also intended for use as a resource book for the education of postgraduate scholars in plant breeding and genetics, and for the continuing professional development of plant breeders. For this purpose, boxes are included in the main sections of each chapter that summarize its educational objectives and present the key messages and questions that are involved; in addition, there is a final box at the end of each chapter that summarizes its overall learning objectives. As noted in the Contents list, the book is also accompanied by open resource e-learning materials for each chapter.

*E-mail: Vivienne.anthony@syngenta.com

†E-mail: g.persley@doylefoundation.org

## Overview of demand-led plant breeding

There are three principles that drive success in demand-led breeding: (i) a target-driven approach; (ii) a demand-led variety development strategy; and (iii) performance indicators to measure progress towards the adoption and widespread use of new plant varieties.

**1. Target-driven approach.** Demand-led breeding is target driven. Emphasis is placed on quantitative goal and target setting in order to enable improved varieties to reach the clients for whom they are designed and to fulfil client expectations. In demand-led breeding, this target-driven approach is exemplified by the following best practices:

- **Variety design:** a detailed list of traits with quantified levels of performance is defined to enable comparison with existing varieties before line progression can take place.
- **Client quantification:** numbers of farmers, their locations, market segments and targeted clients in value chains are quantified at the outset of the breeding project.
- **Variety adoption:** target levels are set for adoption by farmers and monitored for success. Variety registration is important to enable farmers to access a new variety.
- **Development stage plan:** a time plan of activities to generate the data required to make line progression decisions is created before the start of the breeding project. The timing of inputs by clients and managers in making these decisions is determined as part of the stage plan, and the stage gates that are the critical decision points are identified.

**2. Demand-led variety development strategy.** A demand-led variety development strategy is designed for each new variety and includes all of the 'what', 'why', 'who', 'when' and 'how' components. The strategy contains a stage plan for line progression decisions, together with a set of development activities and an investment plan for delivery. Monitoring, evaluation and learning (M&E&L) is an integral part of the project delivery plan. A set of key performance indicators (KPIs) are included in the strategy as targets for evaluation. This strategy and its components are used as the baseline for all M&E&L work.

The quality of the strategy is determined by: (i) visioning, foresight and investigative market research to identify new market opportunities and client demand; (ii) engagement with clients to seek their feedback at key decision points in the stage plan – from new variety design through to post-release variety impact evaluation; (iii) ensuring policy coherence and alignment with the country's national priorities and the enabling policy environment; (iv) realism in determining the costs, benefits and appropriateness of the investment in the breeding programme; and (v) well-designed, technically feasible operational plans for the creation and delivery of each new variety that serves clients in particular market segments and/or agro-ecological zones.

**3. Performance indicators to measure success.** The level of engagement and emphasis placed on the views of clients on the performance and use of new varieties is much higher in demand-led breeding than in other breeding approaches. The success of a new variety and its KPIs are determined by the opinions on, demand for and use of the new variety by farmers and clients within the crop value chains.

Successful demand-led breeding programmes satisfy end-user demand and are highly dependent on the assumptions formed during investigative research and collaboration with clients and stakeholders along the value chain. These assumptions form the strategic pillars for monitoring, evaluation and learning during the development, release and adoption of new varieties by farmers. The required engagement with clients and value-chain stakeholders, and the creation and delivery of a variety development strategy and stage plan will require discussion and approval by senior management. Ultimately, a breeding project should be evaluated in terms of:

- **Meeting trait performance targets:** how close is the performance of the new variety to the benchmark variety design/specification/targets set at the product profile/concept stage (as determined using visioning/forecasting methods, market research and inputs from clients in the value chain)? Specifically, were the genetic improvements required for each of the traits delivered?
- **Satisfying clients' needs:** does the new variety satisfy clients' needs and market demand? Is it preferred to older varieties? Has it been adopted by the target numbers of farmers for whom it was designed?
- **Impact:** does the variety create the economic, social and environmental impact at the individual, household and community level that was defined in the benefits case that was used to justify the investment in the breeding project. *Ex post* impact can be assessed only several years after varietal release.

*How does demand-led variety development add value to current practices?*

- **Development strategy.** Demand-led breeding takes an integrated approach to new variety development. It requires a comprehensive analysis that asks the following questions. Who are the targeted clients? What are their needs and how may these change? What are the technical and regulatory elements of plant breeding? How long will the development plan take? How will the new variety reach the targeted clients and will their requirements be satisfied?
- **Measuring success.** Success is measured by satisfying the demand encapsulated in the product profile and by feedback from farmers and other clients in the value chain on product performance and variety adoption. This requires a comprehensive strategy involving: the delivery of new variety design and variety creation, registration and release of the variety; client awareness building; seed distribution to farmers; and performance and adoption monitoring.
- **Development stage plan.** Demand-led breeding requires a stage plan to be created with transparent time points and timelines for data review and germplasm progression decisions that involve the participation of key clients in the value chain. This helps to maintain clients' commitment to new

designs, enables joint problem solving, manages expectations and stimulates demand for the new varieties.

- **Development planning.** Demand-led breeding creates more complexity because of the broader range of client involvement, trait targets and performance testing. Therefore, to counteract potential delays, greater emphasis is placed on the development by breeders of professional planning skills, as well as an understanding of critical paths and risk mitigation strategies.
- **Participatory breeding.** Demand-led breeding includes, but goes beyond, farmer participatory breeding. It puts more emphasis on regularly consulting and understanding the needs and preferences of all clients and stakeholders in a crop value chain. This involves seeking information from farmers and consumers in both rural and urban areas through participatory appraisal methods.
- **Consultative processes.** Consultation is a continuous requirement throughout the whole of the variety development process, registration and launch, so that a new variety not only supports farmers' requirements for crop productivity and sufficient food for home consumption, but also ensures that production surpluses can enter markets. A development stage plan that includes developing shared ideas and joint decision making with stakeholders in the value chain is critical for success.
- **Variety design and benchmarking.** Demand-led breeding places more emphasis on the systematic, quantitative assessment of varietal characteristics and on creating product profiles with benchmarks for varietal performance and line progression. Consumer-demanded traits are also given more importance. Variety design requires prioritization among the many traits desired by farmers, processors, seed distributors, transporters, retailers and consumers.
- **Registration standards.** Early contact with registration officials is required at the variety design phase, well before a potential new variety is ready to enter official registration trials. Thus, at an early stage, there is need to validate designs, agree standards for consumer-based traits and create interest in the new variety by officials, as these may accelerate the timelines to delivery of the demand-led varieties.
- **Benefits and business cases for investment.** Greater emphasis is placed on analysing and creating compelling business cases for new variety development. This is done by identifying and communicating the full breadth of the quantitative and qualitative economic, social and environmental benefits for clients and stakeholders that are likely to arise from investing in demand-led plant breeding programmes.

## Key Elements of Demand-led Plant Variety Design

### Principles of demand-led plant variety design

Chapter 1 (by Paul Kimani) discusses the status and challenges of agriculture in Africa and ways of transforming the agricultural sector into a modern, commercially oriented sector within the countries of Africa. The chapter reviews the adoption of new plant varieties; on average, there has been 35% adoption of modern varieties for many food crops across sub-Saharan Africa over the past 15 years (Walker and Alwang, 2015). This contrasts with about 60% adoption of new varieties in Asia and 80% in South America. The low adoption rates in Africa are attributed to a range of micro- and macro-level factors, including the availability of seed and access to credit, as well as acceptance by farmers and consumers.

The chapter identifies the key principles of demand-led variety design, and discusses how it differs from and complements other approaches, its relationship to innovation systems and value chains, the role of public policy and social dimensions in demand-led design, and the benefits and risks of implementing a demand-led breeding programme. It also defines the role of the plant breeder and the rationale for breeders to adopt more demand-led variety design, including in setting breeding goals, trait trade-offs and measuring progress.

*Demand-led approaches – ten key points*

- **Understanding clients.** Understanding clients is central to demand-led variety design and increasing the adoption of new varieties: Clarity is required on: who the clients are, what factors influence their buying decisions, and the needs, preferences and problems of each client.
- **Farmer adoption.** Demand-led approaches should increase the likelihood of new varieties being adopted by farmers.
- **Value chains.** Demand-led approaches build on and go beyond farmer participatory breeding. They include consultations not only with farmers but with all clients and stakeholders along the whole crop value chain.
- **Urban and rural consumers.** Breeders must consider needs and preferences of consumers living in both rural and urban environments. Rapid (rural and urban) appraisals can be extended to gathering information not only from farmers but also from consumers and clients who live in towns and cities.
- **Markets and client segmentation.** Breeders need to understand markets and client segmentation to be able to prioritize their breeding targets.
- **Market research and intelligence gathering.** Market research at the start of a breeding programme needs to be complemented with continuing consultations with stakeholders at key decision points along the development stage plan from new variety design to post-market release.
- **Breeding entrepreneurship.** This can contribute to economic growth, better livelihoods for smallholder farmers and increased food security. Improved varieties can change lives.
- **Market creation.** To maximize market creation and nurture innovation, a balance is required between using demand-led approaches and enabling new technologies to drive innovations. Both approaches have value and they complement one another.
- **Role of the plant breeder.** Plant breeders do much more than making crosses and leading selection programmes. A breeder must also be an integrator of inputs and be able to assimilate information and incorporate a broad range of views, including those of non-technical experts. This requires assimilating data, looking at its implications and making decisions based on information from diverse areas, such agricultural economics,

markets and market research, as well as from the core scientific functions for breeding.
- **Breeding experience.** Demand-led approaches retain emphasis and place value on the breeders' eyes and experience in assessing germplasm.

## Visioning and foresight for setting breeding goals

Chapter 2 (by Nasser Yao, Appolinaire Djikeng and Jonathan Shoham) focuses on the skills and methodologies necessary to understand the changes taking place in Africa's food and agricultural production. It describes how to use foresight to anticipate future demand and incorporate these findings into new variety designs. It provides a holistic approach to: (i) analysing the current agricultural landscape and challenges in Africa within a context of market supply and demand; (ii) understanding the drivers of change and their predictability; and (iii) using the methodology of social, technological, economic, environmental and policy drivers (STEEP analysis) and risk mitigation to create scenarios and validate new variety designs.

*How does demand-led variety design add value to current breeding practices?*

- **Future demand.** Demand-led variety design focuses on understanding market requirements and predicting demand in the 5–10 year period after a new variety is released.
- **Visioning and forecasting.** Best practices for the visioning and forecasting that is applicable to demand-led variety design offer new approaches to add value to current postgraduate and professional development programmes for plant breeders.
- **Risk analysis.** Risk analysis considers the uncertainty of future scenarios and the effect that drivers of change can have on future demand.

*Forecasting future landscapes*

- **Changing demand over time.** Foresight analysis is needed to assess for whom the variety is being designed and whether the clients' needs and preferences will change over the projected timetable for varietal release to farmers.
- **Predicting the future.** The use of STEEP driver analysis and scenario-based methods can help to predict the future better, avoid creating redundant varieties and build confidence in plant breeding programmes among investors, governments and R&D managers.

*Integrating foresight into new variety design*

- **Best practices.** Foresight methods are used to review existing variety designs that are being developed and also as a starting point for the creation of new designs. Both approaches are valid. Every trait characteristic in each product profile should be analysed and a decision taken on whether the trait and its benchmark are likely to remain relevant for the intended users over the time required for variety development.

- **Risk management.** Risk analysis and mitigation is an essential procedure for testing the long-term viability of demand-led designs. Decision points are required in the stage plan and the spreading of risk needs to be considered (e.g. understanding the benefits and costs of maintaining many biologically diverse germplasm lines).

## Understanding clients' needs

Chapter 3 (by Pangirayi Tongoona, Agyemang Danquah and Eric Danquah) enables breeders to: (i) define clients and stakeholders; (ii) understand the various categories of clients (including seed distributors, farmers, processors, traders, retailers, marketers and consumers) and their activities in value chains; (iii) identify market segments and their importance in determining the number of new varieties required; and (iv) understand the different types and methods of market research and the best practices for obtaining the information required to design new, 'fit-for-purpose' varieties from clients and stakeholders.

*How does demand-led breeding add value to current breeding practices?*

- **Client focus.** Breeding goals and objectives are set based on what clients want and need without bias towards either what technology can offer or a specific focus on individual trait improvement.
- **Value chains.** Greater understanding is required about the structure of crop value chains, and the buying and selling factors of different clients and their relative priority, when setting new variety designs.
- **Dual-purpose varieties.** A new variety not only supports the farmers' requirements for crop productivity and home consumption, but also ensures that surplus crop production can enter markets with cash returns to all of the value chain participants.
- **Market research.** Stronger emphasis is given to gathering unbiased, reliable and independent information on clients' needs and preferences.
- **Market and business knowledge.** Breeders require greater knowledge about crop uses, markets and the 'business/economics' of breeding.

*Clients within value chains*

- **Understanding clients.** This is central to demand-led variety design, release and adoption. It is essential to be clear on who the clients are and what affects their buying decisions.
- **Value chains.** Breeders need to understand value chains and the relative importance of different clients in the chain and their requirements within each new variety design.
- **Different clients.** Value chain clients have different requirements and not all of these requirements can always be satisfied with the same variety, especially when there are specialist properties required for processing. Breeders should have regular contact with clients in all parts of the value chain and involve them in new variety design.

- **Client location and scale.** The geographic location of clients is important, as is the question of whether the benefits and values of new varieties are also applicable for potential clients across national borders. The analysis of agro-ecological zones should be given particular attention. The more clients that can benefit from each new variety, especially when it can have multi-country impact, the better the investment case for a breeding programme.
- **Seed system development.** For the development of seed systems, i.e. the means by which seeds are produced and obtained, and for improved seeds to reach farmers, especially in remote locations, distributors require portfolios of 'fit-for-purpose' varieties. Portfolios of new varieties are also required for market creation, growth and business sustainability.
- **Public and private sector roles in seed supply.** Public sector breeding programmes are the initial source of new varieties to serve clients and value chains with food security crops that are currently not commercial (export) crops. In the longer term, developing the local private sector seed business is a more sustainable strategy for both food security crops and commercial (export) crops.

## New variety design and product profiling

Chapter 4 (by Shimelis Hussein) aims to enable participants to design new crop varieties that will achieve high adoption rates because their varietal characteristics serve the needs and preferences of farmers, processors, consumers and other stakeholders in the crop value chain.

*How does demand-led variety design add value to current breeding practices?*

- **Variety design and benchmarking.** Stronger emphasis is placed on the systematic quantitative assessment of varietal characteristics and the creation of product profiles with benchmarks for varietal performance and line progression. Consumer-demanded traits are recognized as being as important as production traits. This requires a greater strategic prioritization of traits among the many traits that are required by farmers, processors, seed distributors, transporters, retailers and consumers. This may involve the development of different varieties for different segments of the value chain.
- **Competitor product profiling.** This requires analysis of the characteristics of current commercial varieties and landraces as grown by farmers, and of their differentiating characteristics at every stage in the value chain from seed production to farmers, processors, transporters, retailers, food companies and consumers.
- **New variety design.** A detailed product profile is created that contains many traits and characteristics (typically more than 40) with performance benchmarks that are used to create breeding objectives. Current practices often focus on a much smaller number of farmer requirements that are well understood, but are not discussed or agreed with other stakeholders in the

value chain. Demand-led approaches put more emphasis on combining consumer-based traits with farmer requirements to drive adoption.
- **Quantitative benchmarks.** For each trait, a target quantitative benchmark is set for line progression for variety release, rather than the common procedure of deciding on a defined number of years for annual selection and progressing the best performing lines at the end of the term for registration.
- **Trade-off decisions.** A decision-making process is used that takes into account client needs, technical feasibility and a range of other practical and fiscal considerations. Active and inclusive decision making is core to demand-led breeding. A prioritized list of traits and of the final new variety design that is used to set the breeding goals is discussed and agreed with clients and stakeholders before breeding work commences.

*Variety design*
- **Product profile.** A specific product profile is required for each segment of clients that a new variety is intended to serve. Each product profile comprises a defined set of prioritized traits.
- **Communication.** A consistent format should be used for product profiles so that they are easy to compare and communicate to clients, plant breeders, scientists, managers and other stakeholders.
- **Validation.** Each new product profile should be tested with clients and the assumptions that have been made about acceptability validated before major investment is made in a breeding programme.
- **Market research data.** Qualitative and quantitative data from early discussions with farmers and clients in the crop value chain should be used to create product profiles and make decisions on breeding objectives.
- **Adoption tracking.** Breeders should consider at the variety design stage how adoption tracking will be done (e.g. phenotypic versus genotypic markers) and build these markers into the variety design.
- **Breeding goals.** Validated product profiles that comprise a predefined, integrated and prioritized set of traits should drive the setting of breeding goals and objectives, rather than single traits.
- **Forecasting requirements.** Breeders need to decide how long it will take to develop their new variety and then use scenario-based techniques to review the applicability of their designs on this time frame.

*Setting standards*
- **Breeding objectives.** Clear, quantified breeding objectives with performance indicators are essential.
- **Benchmarks.** Each trait in a product profile should be quantified and measurable versus a defined performance benchmark that needs to be achieved to ensure registration and future adoption by farmers, and based on the performance of a popular variety or landrace.
- **Bioassays.** Performance must be measurable with 'fit- for-purpose' assays.
- **Variety registration requirements.** This process must be understood at the design phase and early discussions held with officials, particularly when

the design includes consumer-based traits, markers for variety identification and the monitoring of trait performance assessment (e.g. nutrition, seed certification).

- **Seed production and scaling.** Key design parameters are how easily seed multiplication can be scaled and the size of the associated costs. These factors need to be taken into consideration at the variety design stage so that the future demand for seed can be satisfied. Seed production costs can make the difference between a variety being commercially viable or not.

## Variety development strategy and stage plan

Chapter 5 (by Rowland Chirwa) addresses the following five issues in variety strategy development and stage planning:

- **New variety development strategy.** A new variety development strategy requires the ability to create a *de novo* demand-led strategy as a key communication document.
- **The development stage plan.** Stage planning and decision making need a clear understanding of the key components and benefits of a demand-led development plan that contains critical decision points (stage gates) and lists the information needed for line progression. The stage plan includes all of the required activities and timelines to create new varieties of the target crop, together with clarity on by whom, when and how decisions will be taken on line progression.
- **Timelines and critical paths.** These entail an understanding of the value of organizing the activities required for developing demand-led varieties into an optimized plan and of how to determine the critical path sequence and conduct critical path analysis.
- **Risk management.** The implementation of risk mitigation measures reduces the likelihood of delays and ensures that outputs are delivered on time.
- **Variety registration.** Registration necessitates an understanding the requirements and timescale required to register a new improved variety for a crop in a country or region, and an engagement with variety registration officials to ensure that registration procedures are able to address market- and consumer-demanded traits.

## Monitoring, evaluation and learning

Chapter 6 (by Jean Claude Rubyogo and Ivan Rwomushana) aims to enable breeders to design, integrate and implement plans that demonstrate best practices in M&E&L in their demand-led breeding programmes, including setting targets based on KPIs. It encourages breeders to reflect on what they consider success will look like, in terms of both their demand-led breeding programme and their own professional performance. It focuses on the core principles of demand-led variety design and best practices in M&E&L, and involves clients in the demand-led process and the setting of KPIs. It also covers the importance, challenges and methods for post-release monitoring of the adoption of new varieties by farmers and other value-chain clients.

*Performance benchmarking*

- **Monitoring and evaluation.** To be successful, demand-led breeding projects require the implementation of best practices in project management, including planning, monitoring, evaluation and learning, from the new variety design and project initiation stage through to variety release, widespread use and the eventual discontinuation of the variety.
- **Demand-led strategy and stage plan.** The strategy and stage plan for each new variety design provides the framework, targets, plan and assumptions for all M&E&L activities. Here, the stage gates provide the review points for evaluation and learning.
- **Clients and stakeholders.** Engaging key clients in the value chain in the formation of the development strategy and the M&E&L process is essential. Specifically, in new variety development projects, key clients should be consulted and involved in decisions at the following stage gates: (i) the decision to invest in the new plant breeding project; (ii) the choice of lead lines to be developed and scaled up; and (iii) the release of the new variety. This engagement will increase the ownership of new varieties and ensure longevity of demand.
- **Key performance indicators.** Purposely tailored KPIs should be included in the development strategy to support and encourage the delivery of demand-led plant breeding goals and objectives. Institutional and breeder performance measures may vary and any conflict of interests should be resolved at an early stage with the institutions' research leadership and management.

*Variety adoption and performance tracking*

- **Rationale and benefits.** Variety adoption and performance tracking enables an understanding of the significance of variety adoption assessment and its links to breeding programmes.
- **Responsibility and funding.** The adoption of new varieties needs to be assessed after their release, and this involves both determining who is responsible for tracking adoption and whether the breeding programme will require additional funding to enable this post-release tracking.
- **Methods and technology.** The use of phenotypic markers or of other low-cost, modern molecular technology should be considered to enable the more effective tracking of new varieties after release.
- **Variety adoption tracking.** Government officials and investors need to support variety adoption tracking with additional finance, resources, best practices and transparency, and the encouragement of plant breeders and their clients in the value chain, as such tracking is a means to improve the future performance of the breeding programme.

## The business case for new variety development

Chapter 7 (by Rowland Chirwa) describes the elements necessary for plant breeders to be able to create a compelling case for investment in demand-led plant breeding to put to research and development (R&D) managers, government officials and financial investors. This includes identifying the benefits and intended beneficiaries of a proposed new breeding programme or project, understanding the principles of return on investment, and clarifying whether the investment in demand-led breeding can be justified in terms of the likely economic, social and environmental benefits versus the costs of developing a new variety.

*Benefits and investment cases*

Greater emphasis is placed on analysing and creating compelling business cases, by identifying and communicating the full breadth of quantitative and qualitative economic, social and environmental benefits that will become available for clients and stakeholders by investing in the proposed demand-led plant breeding programme. The critical issues are:

- **Creating a compelling business case.** It is critical to understand the clients to be served by a plant breeding programme as the basis for creating new varieties that have benefits for all of the clients in the value chain and that deliver an attractive return on investment. A broader and deeper understanding of the range of costs necessary to develop demand-led varieties is also required. These are the essential elements to create business investment cases that are persuasive to government officials, to private and public investors and to other stakeholders in order to secure and retain support for a demand-led breeding programme.
- **Making a clear investment case.** Clarity is required on the rationale and justification for proceeding with a demand-led breeding programme. Investment cases are always assumption based. The quality of the case comes from detailed analysis of the benefits and performance assumptions, including questioning their probability and understanding their sensitivity to factors such as the level of farmer adoption, choice of varieties available and changing variety development costs.
- **Communicating an investment case.** Creating a compelling investment case that is understandable and persuasive to government officials, investors and stakeholders is critical to be able to secure and retain support for a demand-led breeding programme.
- **Return on investment.** Governments, R&D managers and investors need to appreciate that breeding programmes can provide a return on investment rather than the investment being seen only as a budget cost. Managers need to encourage an investment decision-making culture rather than a budget spending culture within breeding programmes. This can be achieved by tracking the adoption of and benefits that accrue from new varieties rather than just monitoring the number of varieties that have been developed by the breeding teams and are registered for release.
- **The business of plant breeding.** Demand-led breeding combines the best practices in market-led new variety design with innovative breeding methods and integrates these with the best practices in business.

## Reference

Walker, T. and Alwang, J. (eds) (2015) *Crop Improvement, Adoption and Impact of Improved Varieties in Food Crops in Sub-Saharan Africa*. CGIAR Consortium of International Agricultural Research Centers, Montpellier, France and CAB International, Wallingford, UK.

# 1 Principles of Demand-led Plant Variety Design

PAUL M. KIMANI*

*Plant Breeding and Biotechnology Program, Department of Plant Science and Crop Protection, University of Nairobi, Nairobi, Kenya*

## Executive Summary and Key Messages

### Objectives

**1.** To understand the current status and challenges facing African agriculture.
**2.** To review modern variety adoption in Africa.
**3.** To understand the principles of demand-led plant variety design and how this approach is similar to and different from current breeding practices.

This chapter discusses the status and challenges of agriculture in Africa and ways of transforming the agricultural sector into a modern, commercially oriented sector within the countries of Africa. It reviews the adoption of new plant varieties, where there has been about 35% adoption of modern varieties for many food crops across sub-Saharan Africa over the past 15 years Table 1.1. This contrasts with about 60% adoption of new varieties in Asia and 80% in South America. The low adoption rates in Africa are attributed to a range of micro- and macro-level factors, including the availability of seed and access to credit, and acceptance by farmers and consumers. The chapter also describes the advantages and disadvantages of the methods used to measure varietal adoption. Further, it identifies the key principles of demand-led variety design: how this differs from and complements other approaches; its relationship to innovation systems and value chains; the role of public policy and social dimensions; and the benefits and risks of implementing a demand-led breeding programme. The chapter also defines the role of the plant breeder and the rationale for breeders to adopt more demand-led variety design, including the setting of breeding goals, consideration of trait trade-offs and measurement of progress.

*E-mail: pmkimani@uonbi.ac.ke

## How is demand-led variety development different from current practices?

- **Broader client focus.** Demand-led approaches put farmers, other clients and consumers at the forefront of new variety design and development.
- **Value chain.** Demand-led approaches build on and go beyond farmer-participatory breeding. They include consultation on needs, not only with farmers, but also with all other clients across the entire crop value chain who are making buying and selling decisions about crops and their products, as well as with the stakeholders who determine the enabling environment in which the business of plant breeding and crop production takes place.
- **Markets and drivers.** Plant breeders need to understand markets and client segmentation to be able to prioritize their breeding targets. Current practices tend to put more emphasis on technology-based approaches and achieving numbers of varietal registrations, rather than focusing on market drivers, including partnering with the private sector to ensure that new, marketable varieties reach farmers.

*Implications for the role of the plant breeder*

- **Leadership role.** The breeder is the main actor in demand-led variety design and carries the responsibility for coordinating, facilitating and linking actors and audiences with diverse interests. Success is highly dependent on the breeder championing a demand-led approach.
- **Building expertise.** Plant breeders will need to learn new skills, especially in the business domain, and work with a range of non-traditional allies for the success of their programmes. They will also need to train and mentor a new generation of young breeders in demand-led breeding approaches.

## Key messages for plant breeders

*Outlook for African agriculture*

- **Key challenges.** Breeders need to understand the current supply and demand challenges facing African agriculture.
- **Policy and science agenda.** Breeders, including postgraduate students, need to understand how demand-led approaches fit with Africa's science agenda, government policy and research and development (R&D) investment plans in their own country and region.

*Modern variety adoption in Africa*

- **Limited adoption.** There is limited adoption of many registered varieties of food crops in Africa. The application of demand-led approaches to variety design can improve adoption levels, and the private sector seed industry succeeds or fails based on designing varieties that clients want and need.
- **Variety design.** This is an important component that affects adoption, along with the availability and affordability of seeds, farmer awareness and risk perception.

- **Measuring adoption levels.** Several methods can be used but all have advantages and disadvantages. The overriding issue is data quality and reliability. Data accuracy can be significantly improved if breeders incorporate phenotypic and genotypic identity markers into their variety designs.

*Breeding goals*

- **Customer preference.** This is a significant factor in the adoption of new varieties. The core goals of a breeding programme should be based on what consumers want and the type of products that will improve the livelihoods of farmers.
- **Breeding performance and tracking adoption rates.** A breeder's goals, incentives and rewards should go beyond the number of new varieties registered and also include the extent of adoption of new varieties by farmers and the performance of these varieties in the field. Adoption monitoring methods should be fully understood, so that appropriate monitoring can be built into a post-registration plan to monitor adoption and performance of new varieties, with attribution of success.

*Demand-led approaches – ten key points*

- **Understanding clients.** Understanding clients is central to demand-led variety design and improving the adoption of new varieties. Before starting a breeding programme for a particular crop, it is necessary to be clear on: who the clients are; what factors influence their buying decisions; and what the needs, preferences and problems of each client are.
- **Farmer adoption.** Demand-led approaches increase the likelihood of new varieties being adopted by farmers.
- **Value chains.** Demand-led approaches build on and go beyond farmer-participatory breeding. They include consultation not only with farmers, but also with all clients and stakeholders along the whole crop value chain.
- **Urban and rural consumers.** Breeders must consider the needs and preferences of consumers living in both rural and urban environments. Rapid (rural and urban) appraisals can be extended to gathering information not only from farmers but also from consumers and clients who live in towns and cities.
- **Markets and client segmentation.** Breeders need to understand markets and client segmentation to be able to prioritize their breeding targets.
- **Market research and intelligence gathering.** The gathering of market research at the start of a breeding programme needs to be complemented by regular consultations with stakeholders at key decision points along the development stage plan from new variety design to post-market release. The primary objective is to gain new insights, test assumptions, demonstrate and obtain feedback on new variety/lead germplasm performance and stimulate demand.
- **Breeding entrepreneurship.** This can contribute to economic growth, better livelihoods for farmers and increased food security. Improved varieties can change lives.

- **Market creation.** To maximize market creation and nurture innovation, a balance is required between using demand-led approaches and enabling new technologies to drive innovations. Both approaches have value and they complement one another.
- **Role of the plant breeder.** Plant breeders do more than make crosses and lead selection programmes. A breeder must also be an integrator of inputs and be able to assimilate information and incorporate a broad range of views, including those of non-technical experts. This requires assimilating data, looking at its implications and making decisions based on information from diverse areas, including agricultural economics, markets, market research and the core scientific functions of breeding.
- **Breeding experience.** Demand-led approaches retain emphasis and put value on the breeder's eye and experience.

## Key messages for R&D leaders, government officials and investors

*Outlook for African agriculture*

- **Strategic agriculture and the food security agenda**. Demand-led approaches can contribute to achieving national, regional and pan-African agricultural development and food security targets and priorities.

*Modern variety adoption*

- **Tracking adoption and performance of new varieties.** Breeders should aspire to and be supported by R&D management and investors to monitor the adoption of their varieties by farmers and by the actors along the value chain.
- **Financial support for monitoring variety adoption.** Finance should be included for use after registration as part of the financial investment in breeding programmes in order to measure progress.

*Breeding goals*

- **Clients.** Client needs and preferences should drive the setting of breeding objectives. The incorporation of demand-led approaches into the setting of breeding goals and objectives will contribute to smallholder farmer livelihoods by enabling them to enter markets for the sale of surplus produce. This will increase the success and reputation of the breeder. It will also increase investors' confidence in continued support for the breeding programmes.

*Demand-led approaches*

- **Demand-led versus technology push.** A balance is required between the use of demand-led approaches and technology/innovation push so that market creation is maximized for new varieties.
- **Change management.** Demand-led approaches require significant changes to the way that current breeding goals and objectives are set, new varieties are designed and developed, and their performance is measured. Support is sought from institutional management, government officials and investors

to assist breeders to adopt more demand-led approaches and to find solutions to problems that may arise during the transition to these new approaches.
- **Private sector connections.** Public sector breeders should be encouraged to connect with the private sector along the whole value chain from seed distributors to retailers in order to seek inputs and stimulate demand. This requires that management and government officials create an enabling environment and facilitate a public/private dialogue.
- **Market and crop information.** Governments and management need to provide reliable national statistics and data sources for breeders and other R&D scientists to use.
- **Market research.** Investors should encourage, expect and finance market information gathering as a key part of breeding programme proposals and business plans.
- **Professional development of plant breeders as integrators.** R&D leaders need to support and finance breeders to undertake professional development in knowledge integration and to expand their skills in project management and project planning.

## Introduction

The objectives of this chapter are:

**1.** To understand the current status and challenges facing African agriculture.
**2.** To review modern variety adoption in Africa.
**3.** To understand the principles of demand-led plant variety design and how this approach is similar to and different from current breeding practices.

The chapter discusses the status and challenges of agriculture in Africa and ways of transforming the agricultural sector into a modern, commercially oriented sector within the countries of Africa. It reviews the adoption of new plant varieties, where there has been about 35% adoption of modern varieties in many food crops across sub-Saharan Africa over the past 15 years (Walker *et al.*, 2014; Walker and Alwang, 2015). This contrasts with about 60% adoption of new varieties in Asia and 80% in South America. The low adoption rates in Africa are attributed to a range of micro- and macro-level factors, including the availability of seed and access to credit, and acceptance by farmers and consumers. The chapter also describes the advantages and disadvantages of the methods used to measure varietal adoption. Further, it identifies the key principles of demand-led variety design: how this differs from and complements other approaches; its relationship to innovation systems and value chains; the role of public policy and social dimensions; and the benefits and risks of implementing a demand-led breeding programme. The chapter also defines the role of the plant breeder and the rationale for breeders to adopt more demand-led variety design, including in the setting of breeding goals, consideration of trait trade-offs and measurement of progress.

The aim of the chapter is to enable African plant breeders to understand the principles and share the best practices of demand-led variety design, and to act as a resource for education in this field. For this purpose, boxes are included in the main sections of the chapter that summarize its educational objectives and present the key messages and questions that are involved. There is also a final box at the end of the chapter that summarizes the overall learning objectives.

## The Transformation of Agriculture in Africa (Box 1.1)

Agriculture is the principal source of livelihood for millions of people living in rural and urban communities across Africa. It is the main determinant of food and nutritional security, employment and incomes, and the prime driver of economic growth. Consequently, poor performance in the agricultural sector affects the performance of nearly all other economic sectors. Agriculture employs over 60% of the workforce and accounts for more than a quarter of the continent's gross domestic product. However, agricultural productivity across Africa lags behind that of most other regions of the world. Africa's global competitiveness, as measured by its global share of agricultural exports, has fallen from about 8% in the 1970s to less than 2% by 2013. Low agricultural productivity leads to food and nutritional insecurity. Productivity and value addition is low across most sectors, making African economies less competitive globally. The lack of economic diversification and inability to create new competitive sectors not only threatens sustainable growth, but also creates challenges in employing existing and new entrants into labour markets and taking people out of poverty. In Mozambique and Tanzania, for example, 500,000 to 800,000 young people enter the labour market each year. The situation is similar in many other countries in sub-Saharan Africa. Hence, Africa requires an economic transformation in the key agricultural sector.

**Box 1.1.** The transformation of agriculture in Africa: educational objectives.

**Purpose:** to familiarize participants with the importance of agricultural productivity within Africa and the effect of population growth on food security, and to introduce the concept of market demand.

**Educational objectives:**

- to understand the current status and challenges facing African agriculture;
- to understand the basis of agricultural transformation in Africa; and
- to appreciate the significance of market-demanded skills and training for graduate students, professional breeders and other agricultural scientists.

**Key questions**

- **Challenges.** What are the current supply and demand challenges facing African agriculture?
- **Policy and science agenda.** How do demand-led approaches fit with Africa's science agenda, government policy and R&D investment plans in each country and region?
- **Strategic agriculture and food security agenda.** How do demand-led approaches contribute towards achieving national, regional and continental targets and priorities for enabling agriculture-driven economic development and ensuring food security?

It is predicted that Africa's population will double to 1.1 billion between 1997 and 2020. As a result, the demand for imported food, mostly cereals and legumes, will increase to 50–70 million t annually. If the current economic situation persists in the food-deficit nations of Africa, then it is unlikely that these nations will have the resources to purchase this huge volume of food on a commercial basis. Several countries have already become regular recipients of food aid, but it is unlikely that the international community will continue to provide food on concessionary terms or as food aid. It is, therefore, necessary for countries in Africa to continue and to accelerate the development and implementation of strategies designed to increase crop productivity and commercialization. It is for this reason that many governments in Africa are aiming to transform their agricultural sectors from subsistence agriculture to market-oriented systems.

The process of this economic transformation requires an increase in science and technology capacity and a skilled workforce to undertake applied research and to accelerate technology absorption and generate new competitive sectors (FARA, 2014). Africa needs local graduates with up-to date skills and knowledge in productive sectors such as agriculture. At present, such graduates are in short supply. Countries need to produce more graduates in science and technology who can both contribute to and benefit from new opportunities as economies grow and diversify, and agricultural markets develop. The gap between labour market demand and the programmes offered by tertiary education institutions has led to a situation where jobs are available but there are not enough suitable graduate students to fill them. To address this gap, actions such as bringing in employer recommendations for curricula, linking with industry and policy makers to set up internships for students, and developing general skills that increase employability – such as learning to learn, problem solving, project work and teamwork, and communications skills – as well as the development of market-demanded products, are critical. The proposed demand-led approaches to plant breeding are designed to address some of these issues for graduate students and professional plant breeders and can contribute to achieving pan-African, subregional and country food security targets and priorities.

## Variety Adoption in Africa (Box 1.2)

Plant breeding is the art and science of improving crop plants for the benefit of people. It is the art and science of changing and improving the heredity of plants. Preferential selection to meet particular human needs results in a broad range of cultivated types within a species. Such cultivated types are also known as varieties or cultivars. When breeders develop new varieties, they expect that they will be utilized by end users such as farmers, consumers and processors. However, not all of the varieties released are utilized by the intended beneficiaries. The acceptance and utilization of new crop varieties is referred to as adoption. The degree of adoption may vary from 0 to 100%, and it varies with crop species and varieties within a species, from one ecological zone to another, and among countries and regions of the world.

A recent study on adoption of new varieties of 20 crop species in 30 countries of Africa over 15 years showed that the average adoption rate of more than 1150 crop varieties developed by CGIAR centres and their partners in Africa is about 35% (Table 1.1) (Walker *et al.*, 2014; Walker and Alwang, 2015). This compares with average adoption rates of new varieties of 60% in Asia and 80% in South America. The study concluded that new maize varieties had the highest adoption, as they are grown in 20 countries in eastern, central, southern and West Africa. New varieties of field pea, which are mainly grown in Ethiopia, had the lowest adoption (1.5%). The adoption of modern varieties was highest in Zimbabwe and lowest in Mozambique.

## Constraints and issues

Many studies have been conducted to determine the reasons for the low adoption of new technologies in Africa. Most have focused on micro factors related to on-farm resources, farmer behaviour, farm market-related factors and variables related to access to services (Kaliba *et al.*, 1998; Tesfaye *et al.*, 2001;

**Table 1.1.** Adoption of modern varieties of food crops in sub-Saharan Africa. Data from Walker *et al.* (2014) and Walker and Alwang (2015).

| Crop | Country observations | Total area (ha) | Adopted area (ha) | % Modern varieties |
|---|---|---|---|---|
| Soybeans | 14 | 1,185,306 | 1,041,923 | 89.7 |
| Maize – WCA[a] | 11 | 9,972,479 | 6,556,762 | 65.7 |
| Wheat | 1 | 1,453,820 | 850,121 | 62.5 |
| Pigeon peas | 3 | 365,901 | 182,452 | 49.9 |
| Maize – ESA[b] | 9 | 14,695,862 | 6,470,405 | 44.0 |
| Cassava | 17 | 11,035,995 | 4,376,237 | 39.7 |
| Rice | 19 | 6,787,043 | 2,582,317 | 38.0 |
| Potatoes | 5 | 615,737 | 211,772 | 34.4 |
| Barley | 2 | 970,720 | 317,597 | 32.7 |
| Yams | 8 | 4,673,300 | 1,409,309 | 30.2 |
| Groundnuts | 10 | 6,356,963 | 1,854,543 | 29.2 |
| Beans[c] | 9 | 2,497,209 | 723,544 | 29.0 |
| Sorghum | 8 | 17,965,926 | 4,927,345 | 27.4 |
| Cowpeas | 18 | 11,471,533 | 3,117,621 | 27.2 |
| Pearl millet | 5 | 14,089,940 | 2,552,121 | 18.1 |
| Chickpeas | 3 | 249,632 | 37,438 | 15.0 |
| Faba beans | 2 | 614,606 | 85,806 | 14.0 |
| Lentils | 1 | 94,946 | 9,874 | 10.4 |
| Sweet potatoes | 5 | 1,478,086 | 102,143 | 6.9 |
| Bananas | 1 | 915,877 | 556,784 | 6.2 |
| Field peas | 1 | 230,749 | 3,461 | 1.5 |
| **Total/rounded weighted average** | **152** | **107,721,630** | **37,969,575** | **35** |

[a]West and Central Africa; [b]East and South Africa; [c]common beans, *Phaseolus vulgaris*.

Abay and Assefa, 2004; Tura *et al.*, 2010). Farmer preferences that act as factors influencing the decision to adopt new varieties have been considered in some studies. Three main, micro-level reasons why farmers do not adopt new technologies were identified by Doss (2005):

- **Awareness.** Simply put, smallholder farmers are either not aware of new technologies or they are unaware that the new technologies would provide benefits for them. Farmers may also have misconceptions about the costs and benefits of the technologies. Negative or positive concepts arise from the 'technological frames' that influence various actors' perceptions and hence their technical choices (Kaplan and Norton, 2008).
- **Availability.** The technologies are either not available or they are unavailable when needed.
- **Profitability.** The technologies are unprofitable given the complex sets of decisions that farmers make about how to allocate their land and labour across agricultural and non-agricultural activities.
- **Gender-based constraints.** These act as a powerful force against the adoption of new technologies. The typically lower asset base of women and so their more limited control over any benefits act as major deterrents to adoption (King and Mason, 2001).

Although these micro-level adoption studies have identified important factors, their macro-level application to spur variety adoption has been limited because they cannot address important political economy issues (Doss, 2005). In the context of agricultural innovation systems (AIS), these adoption studies have focused on the technology (product) per se, with only limited consideration of processes, marketing systems and institutions (Knickel *et al.*, 2008; Mwangi and Kariuki, 2015).

In another study, the Association for Strengthening Agricultural Research in Eastern and Central Africa (ASARECA, a not-for-profit subregional organization of the National Agricultural Research Systems (NARS) of 11 African member countries, which is based in Uganda) sought to establish the macro foundation of the reasons for low rates of technology adoption in eastern and southern Africa (Odame *et al.*, 2013). This study reported that the major macro-level factors responsible for low adoption are:

- soil fertility and agro-ecological targeting;
- seed systems;
- extension services;
- livestock technology delivery mechanisms;
- the performance of the released technologies;
- inadequate attention to gender-based constraints in technology design and delivery;
- lack of commercialization of commodities; and
- overall political economy processes that influence the creation of an enabling policy environment.

In Ethiopia, Zelleke *et al.* (2010) reported that soil-related constraints to adoption and to improved productivity include:

- topsoil erosion;
- acidity (affected soils covering over 40% of Ethiopia);
- significantly depleted organic matter due to widespread use of biomass and dung as fuel;
- depleted macronutrients and micronutrients;
- the destruction of soil physical properties; and
- a rise in salinity.

Without public funding, mitigating such widespread soil degradation in Ethiopia is difficult to deal with because of the uncertain ownership of land – which discourages investment in land development, the low returns to crop and livestock farming, and the challenges arising from variable weather. Thus, Ethiopia has approached the problem as a national concern to ensure that adequate public resources are made available to help stem the decline in soil fertility.

In summary, multiple factors are responsible for low technology adoption, and these can be grouped into micro-level factors and macro-level factors. The main micro-level factors include: those related to farm resources and farmer characteristics (education, age, gender, wealth, farm size, labour, credit, tools, etc.); farming systems (cropping system, soil type and climate); market-related factors, including risk, output market, storage, input market, information; and variables related to access to services (access to credit and membership of cooperatives). Similarly, farmer preferences for technology-specific characteristics as factors that significantly influence the decision to adopt have been considered in some studies. Macro-level factors responsible for low technology adoption include policies, institutions, infrastructure and the dynamics of adoption.

Surprisingly, there are few studies that address specific elements of varietal design and their effects on farmer adoption. Typically, some information is available from *ex ante* impact studies investigating the benefits derived from development interventions. These studies demonstrate that farmer preferences for local varieties are driven by the importance of consumer-based traits, so that varieties with yield improvements and resistance to biotic stresses – but which lack the core consumer requirements – are less preferred. For example, in Kenya, farmers prefer sorghum varieties with desirable consumer attributes such as taste, brewing quality and ease of cooking, and this influences new varietal adoption of sorghum in Kenya (Timu *et al.*, 2014).

## Measuring adoption levels

Several approaches have been used to measure the adoption of new technologies. These include: field surveys using questionnaires; focus group discussions; expert opinion; the use of secondary data from international organizations such as FAO; the use of national statistics from ministries of agriculture and other governmental organizations; reviews of past case studies; and seed sales.

For example, the data from CGIAR's Diffusion and Impact of Improved Varieties in Africa (DIIVA) study (Walker *et al.*, 2014) largely drew on judgements made by expert panels. This remains the dominant method for estimating crop area under modern varieties at a large scale, due to the cost and complexity of collecting data on varietal diffusion through other means. Thus, the DIIVA study relied primarily on expert panel judgments for 115 crop–country combinations. In a number of cases, these expert data were supplemented by estimates based on household surveys (for 36 crop–country combinations).

The DIIVA project was not the first to compare subjective estimates on adoption from expert panels with more objective data. Economists from the International Maize and Wheat Improvement Center (CIMMYT) assessed the congruence between expert opinion from national scientists, mainly plant breeders, and aggregate adoption estimates from data on seed sales of hybrids and open-pollinated varieties (OPVs) for maize-growing countries in southern and East Africa. Their assessment showed that expert opinion on adoption in countries where hybrids were popular and approaching full adoption was very consistent with estimates derived from seed production data. However, divergence between expert opinion and seed sales was higher for OPVs.

**Box 1.2.** Variety adoption in Africa: educational objectives.

**Purpose:** to evaluate modern crop variety adoption in Africa, and to formulate opinions on the importance of post-release monitoring and the best methods to track the use of new varieties by smallholder farmers.

**Educational objectives:**

- to review modern variety adoption in Africa and for a specific region, country and/or crop;
- to compare adoption patterns in Africa with those of other regions of the world;
- to provide a practical exercise on measuring adoption and the interpretation of data; and
- to understand successes and failures of adoption.

**Key messages**

- There is limited adoption of many registered varieties in Africa (as evidenced by quantitative data sets from the DIIVA project and national statistics (Walker *et al.*, 2014).
- Varietal design is an important component that affects adoption, along with availability and affordability of quality seed, farmer awareness and risk perception.

**Key questions**

- Why are farmer adoption levels of modern varieties lower in African countries that in any other region of the world?
- What are the main reasons for adoption or non-adoption?
- How can you measure farmer adoption levels?
- What techniques are used to measure adoption levels?
- What are the pros and cons of each method?
- What techniques were used in the CGIAR DIIVA (Diffusion and Impact of Improved Varieties in Africa) and ASARECA (Association for Strengthening Agricultural Research in Eastern and Central Africa) studies on adoption and what is the justification for these methods?
- What methods are being used by your institute, breeding programme or country to track adoption?
- Which methods are the most appropriate to use for your own breeding programme?
- Can you recognize your varieties in the field, and, if not, why not?
- What actions can you take to be able to recognize your varieties?

In the ASARECA study on technology adoption (Odame *et al.*, 2013) primary adoption data were generated through key informant interviews and focus group discussions. The key informant interviews were conducted with national policy makers, while the focus groups brought together representatives of given enterprise/commodity value chains, e.g. farmers, non-governmental organizations (NGOs), community-based organizations, agro-dealers/stockists, extension workers, agro-processors, wholesale and retail traders, researchers and financial institutions. Focus group participants used the analysis of strengths, weaknesses, opportunities and threats (SWOT analysis) to discuss the adoption of available technologies along a given enterprise value chain. The ASARECA team also reviewed past adoption studies to prioritize both low and high adoption rates, which allowed identification of the successful and missing elements of technology systems.

## Advantages and disadvantages of methods of adoption measurement

Each method of adoption measurement has strong points as well as weaknesses. The overriding issue is the quality of the data generated. The second issue is scale and the extrapolation of localized studies to national, subregional and regional levels. Because of its availability and global coverage, data from the Food and Agriculture Organization of the United Nations (FAO) is widely used despite concerns about its quality. The DIIVA project also reconfirmed the need for field measurement in cases where varieties are difficult to distinguish morphologically (Walker *et al.*, 2014). The survey of cassava in south-western Nigeria epitomizes this case. Farmers knew improved varieties by a group name but could not distinguish relatively small morphological and phenotypic differences between them that allowed for the elicitation of reliable data on specific modern cultivars. In this case, there is no substitute for field measurement, which is more feasible in cassava because it is in the field for a longer time than other crops in a mature state.

Survey performance could be improved if focus groups generated reliable information on varietal adoption. The use of focus group interviews in a community questionnaire was one of the features of the surveys supported by the DIIVA project (Walker *et al.*, 2014). In their validation reports, DIIVA project participants formally compared responses from focus groups and household questionnaires. These reports suggest that focus groups can provide useful information about the relative importance of the variety in the village and the adoption levels by individual farmers; but household data are strongly preferred if cultivar-specific area estimation is the goal. Another issue is the extent to which sampled areas represent reality on the ground. The DIIVA project suggested that well-structured, community focus group discussions combined with field visits could be one such cost-effective alternative to the 500–700 representative household surveys made by the study in order to validate expert opinion from the more qualitative perspective of 'Do the elicited estimates reflect reality or not?'

## Breeding Goals and Objectives

### Setting breeding goals and objectives (Box 1.3)

One of the main activities in planning a breeding programme is to determine the objectives. There are well-established, broad objectives that cut across most programmes. These include yield enhancement, incorporating resistance to biotic and abiotic stresses, adaptation to photoperiod sensitivity, suitability for mechanization and nutritional quality. These breeding objectives are often

**Box 1.3.** Setting breeding goals and objectives: educational objectives.

**Purpose:** to stimulate breeders to reflect on their core mission and on how setting breeding goals and measuring varietal performance against them is central to improving the livelihoods of smallholder farmers, catalysing crop value chains, increasing food and nutritional security in Africa and developing the breeders' own professional reputations.

**Educational objectives:**

- to identify the potential limitations of current breeding approaches;
- to understand the rationale and drivers for incorporating demand-led principles into setting breeding goals and objectives; and
- to be able to develop breeding goals that incorporate client-centric thinking and demand-led approaches into new breeding programmes.

**Key messages**

- Client preference and perception of product performance are critical factors influencing the adoption of new varieties.
- The incorporation of demand-led approaches into breeding programmes will contribute to smallholder farmers' livelihoods and food security by enabling them to enter markets.
- Introducing demand-led principles into goal setting will increase investor confidence in the likelihood of plant breeding success and increases the probability of further investments by investors.

**Key questions**

- What are your breeding goals?
- How do you set them?
- What are the current approaches to setting breeding goals?
- What are the advantages and disadvantages of these goals?
- How do you set targets for your breeding programme?
- In setting your breeding targets, what should be your mission and primary goals?
- Should breeders aim to register varieties with improved trait characteristics so that farmers have choice and/or create varieties that achieve high adoption rates by farmers?
- What are the key factors and trade-offs to consider?
- What does success look like?
- How do you measure progress towards your breeding goals?
- What are the best key performance indicators (KPIs) to use?
- What properties should KPIs have?

derived from local experience with farming communities, national and regional priorities, socio-economic surveys of potential end users and global trends. Breeding goals are often driven by broad based aspirations such as yield gains or drought resistance or are trait specific, with little consideration given to the suitability of the new varieties to meet specific client demands and the incorporation of changing demands to meet new market opportunities. Breeders rarely use market research to determine precisely what their customer demands are and how these demands are likely to change over time. When the client needs are neither well understood nor taken into consideration during the setting of breeding goals and objectives, the resulting new varieties are unlikely to meet consumer needs and often result in slow adoption or even rejection by the intended end users.

#### *Limitations of current approaches*

A recent study by ASARECA on the causes of low adoption of new agricultural technologies in eastern Africa found that the poor performance of these technologies is a major cause of low adoption (Odame *et al.*, 2013). Technology delivery and the targeting of released varieties for the particular agroecologies are limited. For example, some new varieties of maize were found to be highly susceptible to pests and have a low shelf life, which discouraged farmers from adopting them. In another example, the high biomass of some new varieties of sunflower caused high soil nutrient depletion and farmers felt that the change in yields did not compensate for the deterioration in soils and therefore discontinued the production of these varieties (Odame *et al.*, 2013). The ASARECA study also found that consumers in Uganda had strong preferences for the colour and taste of their staple food of *matooke*, which is made from East African cooking bananas. Consumers considered that the new banana cultivars had not incorporated their preferred food taste characteristics for *matooke*. For dessert bananas, new varieties were not favourable when compared with the traditional 'Bogoya' cultivar.

Consumers have called for regular market surveys and the development of supply chains around smallholder farmers, with complementary investments in all links in the supply chain. Empirical findings showed that technologies for commercialized enterprises linked to marketing systems are better adopted than enterprises with poor marketing systems.

#### *Refining breeding goals and objectives to reflect changing preferences and needs*

Several approaches can be used to determine breeding goals and objectives. These include participatory rural appraisal, problem analysis, market research, field surveys, consumption surveys and needs assessment appraisals. It is important that breeders use these tools to ensure that their products meet the ever-changing consumer preferences and needs.

The core goals of a breeding programme should be based on what consumers want and on the type of products that will improve the livelihoods of farmers. Farmers do not only grow varieties to feed their families, but also to generate income to meet their household financial needs. Governments in many countries of Africa have formulated policies intended to transform agriculture from subsistence farming to a commercially oriented activity.

In the context of plant breeding, this agricultural transformation implies that varieties grown by farmers should have consumer-preferred traits and therefore be marketable. It also implies that breeders should assess the traits offered by current commercial varieties and identify the gaps that will be filled by the new varieties. Breeders should also identify the size and location of markets they are targeting. This approach requires a change in how breeders develop breeding objectives and implement breeding plans so that their breeding strategy is demand led, with consumer demands at the centre of choosing the key traits of new varieties.

The key questions a farmer will ask are: 'What traits and advantages do the new varieties offer compared with existing varieties?' and 'Are these changes sufficiently compelling to justify a shift to the new variety?' For example, an important question is how much yield increase is required for farmers to adopt a given technology, especially if they are using credit to purchase new seeds and other inputs. According to Baum *et al.* (1999), the net benefit should usually be between 50 and 100%, which corresponds to a benefit–cost ratio of 1.5:2. If the technology is new to the farmer and requires the learning of new skills, a minimum rate of return nearer to 100% is a reasonable estimate to assume high adoption.

*Trait trade-offs*

Breeders often have to deal with a broad array of traits desired in a new variety. It is challenging to try to incorporate all of the useful traits with the same level of expression into a single 'ideal' variety. The range of potential traits considered is likely to increase when post-production traits such as taste, cooking time, physical appearance and shelf life are also considered in a demand-led breeding design. This implies that some form of trait trade-off, or what breeders refer to as 'index selection', will become necessary. One approach to deal with trade-offs is to categorize and then rank the priority traits. Such categories could be plant traits (including plant architecture), resistance to biotic stresses (pests and diseases), tolerance to abiotic stresses, postharvest traits and utilization traits. Prioritization may first be based on basic traits – those that any cultivar must have for a specific market, secondly, on traits that increase market share and, thirdly, on new traits that are not available in existing cultivars. Priorities within categories can be used to develop a selection 'index' and to determine the breeding goals and objectives. Decision-tree analysis can also be used to aid decision making.

## Measuring progress towards goals (Box 1.4)

Breeding programmes often have a time frame during which set targets and milestones leading to programme goals should be achieved. This is essential because breeding is a multistage process that may take several years, depending on crop species, breeding systems and local regulations governing

**Box 1.4.** Measuring progress towards goals: definitions.

A ***performance indicator*** is a unit of measurement that specifies what is to be measured along a scale or dimension, but does not indicate the direction or change. Performance indicators are a qualitative or quantitative means of measuring an output or outcome with the intention of gauging the performance of a programme or investment. These indicators are neutral; they do not indicate directionality or embed a target.

A ***target*** specifies a particular value for a performance indicator to be accomplished by a specific date in the future. It is important for breeders to establish realistic targets for each performance indicator in relation to the baseline data identified. This sets the expectations for performance over a fixed period of time, compared to a set of conditions existing at the outset of a programme or investment. The results will be measured and assessed as progress towards achieving the target in relation to the baseline data.

***Baseline data*** are collected at one point in time and used as a point of reference. If reliable data on performance indicators exist, then such data should be used. If not, the breeder will have to collect a set of baseline data at the first opportunity. In a breeding programme, baseline data refers to the performance of existing commercial varieties (also known as 'check varieties') or the level of trait expression in specific check varieties against which new varieties will be measured.

the release of new varieties. The main phases in a breeding programme are population development, line development, line testing, validation of candidate varieties and variety release. It is essential to track progress towards the programme goals set at the onset of the breeding process. Progress indicators may be set at the whole-programme level and/or for specific target traits.

For example, the number of populations, number of lines with desired traits, number of candidate varieties and number of varieties are possible targets at the whole-programme level. At the trait-expression level, performance indicators such as number of resistances combined, number of lines with better expression of a trait above the benchmark or expected trait improvement at a particular level (e.g. 90% more iron than a check variety by year X ) can be used. These indicators are used to track progress towards the goal. The indicators must be 'SMART' – simple, measurable, achievable, realistic and time-bound.

Selecting the right indicators is vital to effective project management. In addition to being SMART, indicators should be valid, reliable, sensitive, simple, useful and affordable. Indicators are considered to be:

- **Valid** – if they measure what is expected to be measured.
- **Reliable** – if they are consistent over time and will consistently produce the same data if applied repeatedly to the same situation over time.
- **Sensitive** (i.e. responsive to change) – if an indicator is able to detect small but significant changes over time.
- **Simple** – if it will facilitate easy collection of data and the equipment and/or expertise needed to track the performance indicator is readily available.
- **Useful** – if the information collected can be used for decision making and the performance indicator is expressed in a way that will resonate with the intended audience.

- **Affordable** – if the programme can afford to measure the indicator, given the need for timely, accurate information and the potential cost, it is worth the value of the information obtained.

## Principles of Demand-led Approaches to Plant Variety Design (Boxes 1.5 and 1.6)

Demand-led plant breeding is an approach that enables plant breeders to develop more high-performing varieties that meet client requirements and market demand. The approach is based on six core principles: client needs and preferences; value chains; market research; market trends and drivers; the integration of public and private sector capabilities; and multidisciplinary teams. These principles are defined below, together with a description of their roles in implementing a demand-led approach to breeding.

- **Client needs and preferences.** A clear understanding is needed of the needs and preferences of smallholder farmers, processors, traders, retailers, consumers and other actors along a value chain. This is critical in setting the priorities of the breeding programme.
- **Value chain analysis and innovation systems.** A 'value chain' is a set of value-adding activities performed by all actors from production through to the consumption of a specified product. Analysis of the value chain of the target crop species – and of the agricultural innovation system in which the value chain operates – is a key component of demand-led breeding. Value chain analysis of the target agro-enterprise/crop species is used as an organizing tool to track the stages of the technology, the actors and their roles, and links to technology delivery and use. Value chain analysis helps in understanding the buying and selling decisions of each stakeholder in the supply chain from farmer (or producer) to consumer; it identifies the key buying factors that influence each actor in the value chain.
- **Market research.** Market research is used to define the performance standard and priority of each varietal characteristic. It is also used to test and validate key assumptions throughout the variety development process. Demand-led breeding is based on market research. It makes rigorous use of proven market research tools with farmers, consumers and stakeholders.
- **Market trends and drivers.** Demand-led breeding requires an understanding of market trends and drivers. Longer-term visioning of key drivers of change and of the needs of farmers, markets, national and regional production trends and trade policies, the regulatory environment and the biophysical environment, including predictions on climate change, are key inputs into the design of new varieties and the implementation of breeding programmes. Clear visibility of development timescales is also required.
- **An integrated approach to link public and private sectors.** Demand-led breeding uses both public and private sector expertise and integrates the best practices from both sectors into the variety development process. It involves linking breeders with seed production and distribution systems and value

chains. Best practices are identified and used to conduct market research, in the setting of breeding priorities, in designing variety specifications, in the formulation of technological solutions and in linking farmers with markets. The approach relies on the synergy of public–private sector partnerships to achieve goals, targets, results and benefits which cannot be realized by either party alone.

- **Multidisciplinary teams.** Demand-led varietal design and solution development are conducted using a multidisciplinary team approach. Because demand-led breeding follows an innovation system and a value chain approach, it requires a broad range of competencies and actors with different roles and responsibilities to develop a new variety. Gender and other cross-cutting social dimensions should be well integrated into the generation, delivery and use of new technologies. During the variety development process, the full range of creative possibilities, including feasibility, cost and trade-offs between needs and the likelihood of success, are considered. Finally, proof of concept and prototype testing with farmers, consumers and value chain actors throughout variety development and release/distribution stages are integral parts of demand-led variety design.

## Innovation systems and value chains (Box 1.5)

Innovation systems complement value chain analysis. They go beyond a narrow focus on the attributes of productive technologies to encompass the whole system. The main qualitative elements of innovation systems are the actors concerned and their roles and linkages, and the interactions of producers and other users of technologies and mediating institutions. The main elements within an innovation system are: (i) a knowledge and education domain; (ii) a business and enterprise domain; and (iii) bridging institutions that link the two domains. The knowledge and education domain consists of breeding research and education systems. The business and enterprise domain comprises a set of value chain actors and activities that use outputs from the knowledge and education

**Box 1.5.** Innovation systems: definitions of different types of innovations.

**Product innovation:** a product or service that is new or significantly improved. This includes significant improvements in technical specifications, components and materials, software in the product, user friendliness or other functional characteristics.

**Process innovation:** a new or significantly improved production or delivery method. This includes significant changes in techniques, equipment and/or software.

**Marketing innovation:** a marketing method that involves significant changes in product design or packaging, product placement, product promotion or pricing.

**Organizational innovation:** a new organizational method in business practices, workplace organization or external relations.

domain, and independently innovate. Linking these domains are the bridging institutions, such as extension services, political channels and stakeholder platforms, which facilitate the transfer of knowledge and information between the domains. These domains are supported by an enabling public policy and institutional environment.

**Box 1.6.** Principles of demand-led approaches to plant variety design: educational objectives.

**Purpose:** to understand the core principles of demand-led variety development, how these principles can contribute to higher farmer adoption of improved varieties and the implications of the implementation of demand-led approaches within crop breeding programmes.

**Educational objectives:**

- to understand the core principles, similarities and differences of demand-led breeding versus technology-driven approaches to designing new varieties;
- to analyse and identify best practices in case studies from Africa or other regions and/or on important food crops that illustrate how demand-driven considerations can increase the use of improved varieties;
- to integrate client needs and preferences into variety designs, variety development programmes and timelines; and
- to manage effectively the multifunctional expertise, inputs, handovers and communication about the interdependencies between different experts in the team.

**Key messages**

- Demand-led approaches go beyond farmer participatory breeding to include all the key stakeholders in the value chain. Also, demand-led approaches retain emphasis on the value of the breeder's eye and experience.
- A balance is required between using both demand-led approaches and new technology/innovation approaches to maximize market creation for new varieties.
- The role of the plant breeder is much more than leading crossing or selection programmes. S/he must also be an 'integrator' of inputs and be able to take on board information from a broad range of technical and non-technical experts and clients.
- The breeder must understand that customers need to be central to variety design and be clear on who these customers are and what factors influence their 'buying decisions'.

**Key questions**

- What is demand-led breeding and how is it different from or similar to farmer-participatory breeding?
- What are the benefits and risks of using demand-led approaches (*market pull*) and/or technology/innovation approaches (*technology push*)?
- How do you decide the appropriate balance of market pull versus technology push for your breeding programme?
- What are the key roles and responsibilities of a breeder using demand-led methodology?
- What new areas of expertise and information should you integrate into your thinking?
- Who currently uses demand-led breeding approaches in Africa?
- How should the core principles of demand-led breeding in the private sector be adapted to fit public–private sector partnerships in Africa?
- What are the challenges and disadvantages of using demand-led approaches?

## Role of the breeder

The breeder is the main actor in demand-led breeding. In addition to providing technical expertise, the breeder is expected to be the champion of this demand-led approach, and carries the responsibility of coordinating, facilitating and linking actors and audiences with diverse interests. The breeder will need to learn new skills, especially in the business domain, and work with a range of non-traditional allies for the success of his/her programme. The breeder will also need to train and mentor a new generation of young breeders in demand-led breeding approaches.

## Similarities and differences between demand-led breeding and farmer-participatory breeding

There are similarities and differences between demand-led plant breeding and farmer-participatory plant breeding. Participatory plant breeding (PPB), in its current form, has developed over the past two decades. It can be defined as embodying 'approaches that involve close farmer–researcher collaboration to bring about plant genetic improvement within a species'.

There are similarities between demand-led approaches and PPB. Both approaches contribute towards one or more of six stages within a breeding cycle, namely: setting goals and objectives; generating variability (crosses, landraces, gene bank accessions); selecting experimental varieties; testing experimental varieties; variety release and promotion; and seed multiplication and distribution. Both approaches aim at developing client-specific products and the increased adoption of end products.

However, there are major differences between PPB and demand-led approaches. PPB is a highly localized activity that is focused on local needs and with end products that are designed to suit specific environments. In contrast, demand-led breeding is far broader in scope and targets large areas of agro-ecological zones where the crop can be produced at national, regional or even global levels. The key actors in PPB are the farmer and breeders. In contrast, demand-led approaches target all of the actors in a value chain and innovation system. These actors, who include processors, traders, retailers, agro-input dealers, consumers and policy makers, have different interests.

Demand-led variety design emphasizes markets, their demands and trends, and uses a broad range of tools, including market research, modern product promotion tools and value addition. While demand-led design seeks to combine the best practices from both the public and private sector, PBB focuses on local practices and harnesses the expertise of farmers and breeders. Seed distribution in PPB is limited to the locality of its operations. In contrast, demand-led breeding aims to disseminate the seeds of new varieties widely – nationally, regionally and globally. Demand-led breeding aims to use the best practices for the packaging, labelling, promotion and use of distribution networks, and is largely commercially oriented in terms of generating production in excess of local (family) consumption needs to meet market demands and increase incomes for smallholders through their participation in markets.

### Benefits and risks of demand-led approaches

Demand-led breeding has multiple benefits, including higher adoption rates, the tapping of larger markets and hence the use of economies of scale. It can lead to better returns to investment and is potentially attractive to private investment. It is more sustainable in that the private sector is likely to continue with the production and marketing of seed of new varieties that meet market needs. In the longer term, the public and private sector linkages will be strengthened, based on mutual benefits. Demand-led breeding is likely to make significant and sustainable contributions towards national goals of food and nutrition security. The main risks include failure to meet the targets, unfavourable political environments and the slow adoption of new varieties. Another risk is the continued willingness (or not) of actors in a value chain to collaborate over a sustained period.

### The balance between demand-led breeding and technology-driven approaches

Demand-led breeding focuses on putting clients, markets requirements and value chain needs at the centre of the design and development process for new varieties. The design of new varieties and prioritization of traits for breeding is driven by these needs.

In contrast, technology-driven approaches to variety development tend to have broad-based goals in order to allow exploration of the limits of what science can deliver to solve problems. This is done deliberately in working at the frontiers of science and when new technology is emerging, such as in the early days of the genomics revolution. The approach enables scientific creativity to flow and not to be constrained by the scope of immediate market demands. It also enables paradigm shifts to occur and new markets to emerge. Technology-driven approaches may also be relevant where markets or information do not exist and consequently innovations have to rely more heavily on concept testing and prototyping with potential customers. The challenge for breeders is to achieve a balance between these two different but complementary approaches.

## Learning Methods (Box 1.7)

Before this chapter concludes, a summary is provided in Box 1.7 of learning methods – together with assignments and assessment methods – for use with the main topics that have been covered in the chapter: Variety Adoption in Africa; Setting Breeding Goals and Objectives; and the Principles of Demand-led Breeding.

**Box 1.7.** Learning methods, assignments and assessment methods.

**Variety Adoption in Africa**

*Learning method*

- PowerPoint presentation on adoption levels of modern varieties.
- Review the DIIVA study in Africa (Walker *et al.*, 2014) and the ASARECA study (Odame *et al.*, 2013) and your own country literature sources.
- Group discussion on the reasons why adoption levels of new varieties are lower in Africa than in other regions of the world?
- Group discussion on methods for easy identification of new varieties and sharing best practices.

*Assignment*

- Review the literature and data sources on the adoption levels of varieties of your crop and breeding programme in your country. Explain the situation and the reasons why different varieties are preferred, and the key messages and points of learning you will incorporate into your own designs for your crop breeding programme.
- Preparation of a post-release variety monitoring programme, with objectives, methods and costs required.

*Assessment*

- Assignment.
- Exam questions to test understanding of the different methods for monitoring varietal adoption, with their advantages and disadvantages.

**Setting Breeding Goals and Objectives**

*Learning method*

- PowerPoint presentation on best practices in setting and monitoring breeding goals.
- Group discussion on the core mission of being a professional breeder and using demand-led approaches, debating questions such as:
  - Should breeders aim to: (i) register varieties with improved trait characteristics so that farmers have choice; and/or (ii) create varieties that value chains want and achieve high adoption rates by farmers?
  - How should the professional performance of breeders be measured?
  - Is achieving variety registrations enough or should breeders be assessed and rewarded on whether their varieties are grown by farmers and used in the value chain?
  - What are the skills and qualities required to be an excellent demand-led breeder?

*Assignment*

- Participants to reflect on their current breeding or project research goals and objectives and make an assessment of how demand-led orientated they are.
- Suggestions to be provided on how to make the goals more client focused and on goal rationale.

*Assessment*

- Assignment.
- Exam questions on how to set demand-led goals, objectives and key performance indicators.

*Continued*

**Box 1.7.** Continued.

**Principles of Demand-led Breeding**

*Learning method*

- PowerPoint slides explaining the principles of demand-led plant breeding and comparing it with other approaches.

*Assignment*

- Participants to review their current breeding or project research programme to explain what type of breeding approach they are using and why; and, if they wanted to put more emphasis on creating a demand-led programme, what specific interventions, changes and actions would they implement?

*Assessment*

- Assignment.
- Exam questions to test understanding of the similarities and differences between demand-led, technology-led and farmer-participatory breeding approaches.

## Conclusion

It is timely for demand-led approaches to be incorporated into the education of the next generation of breeders in Africa because of:

- the emergence of growing local, national, regional and global markets for the products of African agriculture;
- the need for improved varieties to contribute to the food supply and demand imbalance in Africa, especially for food security crops; and
- the imperative for greater private sector participation in the business of plant breeding in Africa, both to enable greater smallholder access to the new technologies that are embedded in the seeds of new varieties, and to enable participation by smallholders in expanding markets for crop-based products.

In summary:

- Demand-led approaches aim to make the business of plant breeding in Africa more responsive to market demands.
- Demand-led approaches go beyond farmer-participatory breeding to include all of the key stakeholders in the value chain.
- Demand-led approaches retain emphasis on the value of the breeder's eye and experience.
- An appropriate balance is required between using demand-led approaches and a technology/innovation push to maximize market creation for new varieties.
- The role of the plant breeder in demand-led approaches is much more than leading crossing or selection programmes. S/he must also be an integrator of inputs and be able to take on board information from a broad range of technical experts and clients.

- The breeder must understand that customers need to be central to variety design and to be clear on who these customers are and what factors influence their buying decisions.

## Resource Materials

Slide sets are available for this chapter as part of Appendix 3 of the open-resource e-learning material for the volume. These summarize the chapter contents and provide further information. The e-learning material is available at http://www.cabi.org/openresources/93814 and also on a USB stick that is included with this volume.

## References

Abay, A. and Assefa, A. (2004) The role of education on the adoption of chemical fertiliser under different socioeconomic environments in Ethiopia. *Agricultural Economics* 30, 215–228.

Baum, E., Gyiele, L., Drechsel, P. and Nurah, G.K. (1999) *Tools for the Economic Analysis and Evaluation of On-farm Trials*. IBSRAM Global Tool-kit Series No 1, International Board for Soil Resources Management (IBSRAM), Bangkok.

Doss, C.R. (2005) Analysing technology adoption using micro-studies: limitations, challenges, and opportunities for improvement. *Agricultural Economics* 34, 207–219.

FARA (2014) *Science Agenda for Agriculture in Africa (S3A): "Connecting Science" to Transform Agriculture in Africa*. Forum for Agricultural Research in Africa (FARA), Accra. Available at: http://faraafrica.org/wp-content/uploads/2015/04/English_Science_agenda_for_agr_in_Africa.pdf (accessed 2 May 2017).

Kaliba, A.R.M., Verkuijl, H., Mwangi W., Moshi, A.J., Chilagane, A., Kaswende, J.S. and Anandajayasekeram, P. (1998) *Adoption of Maize Production Technologies in Eastern Tanzania*. International Maize and Wheat Improvement Center (CIMMYT), Texcoco, Mexico with Ministry of Agriculture Research and Training Institute, Ilonga, Tanzania and Southern Africa Centre for Cooperation in Agricultural and Natural Resources Research (SACCAR), Gaborone, Botswana.

Kaplan, R.S. and Norton, D.P. (2008) *The Execution Premium: Linking Strategy to Operations for Competitive Advantage*. Harvard Business School Publishing Corporation, Boston, Massachusetts.

King, E.M. and Mason, A.D. (2001) *Engendering Development Through Gender Equality in Rights, Resources, and Voice*. A World Bank Policy Research Report, [No.] 21776, January 2001. World Bank, Washington, DC.

Knickel, K., Brunori, G., Rand, S. and Proost, J. (2008) Towards a better conceptual framework for innovation processes in agriculture and rural development: from linear models to systemic approaches. In: *Proceedings of the 8th European IFSA Symposium, 6–10 July 2008, Clermont-Ferrand, France*. International Farming Systems Association – Europe Group, Universität für Bodenkultur Wien, Vienna, pp. 883–893.

Mwangi, M. and Kariuki, S. (2015) Factors determining adoption of new agricultural technology by smallholder farmers in developing countries. *Journal of Economics and Sustainable Development* 6, 208–216.

Odame, H., Kimenye, L., Kabutha, C., Dawit, A. and Oduori, L.H. (2013) *Why the Low Adoption of Agricultural Technologies in Eastern and Central Africa?* Association for Strengthening Agricultural Research in Eastern and Central Africa (ASARECA), Entebbe, Uganda.

Tesfaye, Z.B., Tadesse, A. and Tesfaye, S. (2001) *Adoption of High-yielding Maize Growing Regions of Ethiopia*. EARO Research Report No. 41, Ethiopian Agricultural Research Organization (EARO), Addis Ababa.
Timu, A.G., Mulwa, R.M., Okella, J. and Kamau, M. (2014) The role of varietal attributes on adoption of improved seed varieties: the case of sorghum in Kenya. *Agriculture and Food Security* 3(9), 1–7.
Tura, M., Aredo, D., Tsegaye, W., La Rovere, R., Tesfahun, G., Mwangi, W. and Mwabu, G. (2010) Adoption and continued use of improved maize seeds: case study of central Ethiopia. *African Journal of Agricultural Research* 5, 2350–2358.
Walker, T. and Alwang, J. (eds) (2015) *Crop Improvement, Adoption and Impact of Improved Varieties in Food Crops in Sub-Saharan Africa*. CGIAR Independent Science and Partnership Council (ISPC) Secretariat, Rome and CAB International, Wallingford, UK.
Walker, T., Alene, A., Ndjeunga, J., Labarta, R., Yigezu, Y., Diagne, A., Andrade, R., Andriatsitohaina, R.M., de Groote, H., Mausch, K. *et al.* (2014) *Measuring the Effectiveness of Crop Improvement Research in Sub-Saharan Africa from the Perspectives of Varietal Output, Adoption, and Change: 20 Crops, 30 Countries, and 1150 cultivars in Farmers' Fields*. Report of the Standing Panel on Impact Assessment (SPIA), CGIAR Independent Science and Partnership Council (ISPC) Secretariat, Food and Agriculture Organization of the United Nations (FAO), Rome. Available at: http://impact.cgiar.org/sites/default/files//pdf/ISPC_DIIVA_synthesis_report_FINAL.pdf (accessed 15 May 2017).
Zelleke, G., Agegnehu, G., Abera, D. and Rashid, S. (2010) *Fertiliser and Soil Fertility Potential in Ethiopia: Constraints and Opportunities for Enhancing the System*. International Food Policy Research Institute, Washington, DC.

# 2 Visioning and Foresight for Setting Breeding Goals

NASSER YAO,[1]* APPOLINAIRE DJIKENG[2] AND JONATHAN L. SHOHAM[3]

[1]*Biosciences eastern and central Africa (BecA), International Livestock Research Institute (ILRI), Nairobi, Kenya;* [2]*Centre for Tropical Livestock Genetics and Health, The Roslin Institute, The University of Edinburgh, U.K.* [3]*Syngenta Foundation for Sustainable Agriculture, Basel, Switzerland*

## Executive Summary and Key Messages

### Objectives

**1.** To empower plant breeders and leaders in research and development (R&D) to consider future agricultural landscapes across Africa.
**2.** To equip breeders with methodologies to design new varieties that will remain relevant and satisfy market demands over time.
**3.** To identify drivers that may affect whether farmers adopt new varieties in the future.

The chapter focuses on equipping breeders with the skills and methodologies to consider the changes taking place in Africa's food and agricultural production. It shows breeders how to use foresight to anticipate future demand requirements and incorporate these into new variety designs. Specifically, it provides a holistic approach to: (i) analysing the current agricultural landscape and challenges in Africa within a context of market supply and demand; (ii) understanding the drivers of change and their predictability; and (iii) using the methodology of social, technological, economic, environmental and policy drivers (STEEP analysis) and risk mitigation to create scenarios and validate new variety designs.

### How does demand-led variety design add value to current breeding practices?

- **Future demand.** Demand-led variety design focuses on trying to predict demand in the 5–10 year period after a new variety is released.

*Corresponding author: E-mail: n.yao@cgiar.org

  *The Business of Plant Breeding* (Eds G. J. Persley and V. M. Anthony)

- **Visioning and forecasting.** The application of best practices for visioning and forecasting to demand-led variety design offers new approaches to add value to current postgraduate and professional development programmes for plant breeders.
- **Risk analysis.** Risk analysis of demand-led design considers the uncertainty of future scenarios and the effect that drivers of change can have on future demand.

*Implications for role of the plant breeder*

- **Visioning and foresight.** These are key skills for demand-led breeders. Breeders need to be aware of the broader considerations that can affect the supply and demand equation in food production and the livelihoods of smallholder farmers and value chains, and lie beyond the scientific and technical areas of plant breeding and genetics. This requires the gathering and interpretation of a range of information from other disciplines and from a wide variety of sources.
- **Partnering with a wide variety of expertise.** Foresight requires breeders to interact with other specialists, such as economists, social scientists, private sector business managers, environmental and climate change specialists, public officials and policy makers.

## Key messages for plant breeders

*Current trends*

- **Trends.** Breeders need to be aware of the changes taking place in African agriculture in both their own and neighbouring countries, and of the policy landscape for their work.
- **Baseline data.** It is important for breeders to access reputable data sources and to have a baseline of information that will enable them to investigate key trends in their home country.

*Forecasting future landscapes*

- **Timetable assumptions.** It is critical to understand the mostly likely development and registration timetable for each new variety design and to have a rigorous and comprehensive development plan (see also Chapter 5, this volume).
- **Changing demand.** Foresight analysis is needed to determine who the variety is being designed for and whether this client group's needs or preferences will change along the projected timetable for release to farmers. Relying only on current market research information to predict what the future will look like on these long timescales is a high-risk approach.
- **Predicting the future is difficult.** Using driver- and scenario-based methods (STEEP) can help to avoid the creation of redundant varieties and builds confidence in plant breeding programmes among governments, investors and R&D managers.

*Integrating foresight into new variety design*

- **Best practices.** Foresight methods are used to review variety designs that are currently being developed or as a starting point for the creation of new designs. Both approaches are equally valid. Every trait characteristic in each product profile should be analysed and a decision taken on whether the trait and benchmark is likely to remain relevant for its intended users during the time period required for variety development.
- **Risk management.** Risk analysis and mitigation is an essential procedure and a cornerstone of testing the long-term viability of demand-led designs. Decision points are required in the stage plan and risk spreading needs to be considered (e.g. understanding the benefits and costs of maintaining many biologically diverse germplasm lines).

### Key messages for R&D leaders, government officials and investors

*Current trends*

- R&D leaders and managers need to support their staff by understanding and communicating key trends and providing reputable data sources to use for scenario analysis.

*Forecasting future landscapes*

- **Key drivers of change.** R&D leaders should monitor and review the broad range of the key drivers that can change the future agricultural supply and demand landscape, and discuss the implications of the results obtained with their breeders. Incorrect estimates or wrong assumptions can invalidate designs and lead to low varietal adoption.
- **Scenario analysis.** R&D leaders, governments and public and private investors in breeding programmes need to understand the time that it takes to develop new varieties and to support breeders in using the best methodologies to test and review the validity of their breeding goals, objectives and product profiles.
- **Partnerships with experts.** Experts in social trends, advances in core breeding technologies, agricultural economics, climate change, government policy and value chains should be identified and their partnerships with plant breeders encouraged and supported.

*Integrating foresight into new variety design*

- **Annual reviews.** R&D leaders and managers should commission annual reviews of their institution's plant breeding designs and risk-mitigation strategies.

## Introduction

The objectives of this chapter are:

**1.** To empower plant breeders and research and development (R&D) leaders to consider future agricultural landscapes across Africa.

**2.** To equip breeders with methodologies to design new varieties that will remain relevant and satisfy market demands over time.
**3.** To identify drivers that may affect whether farmers adopt new varieties in the future.

The chapter focuses on equipping breeders with the skills and methodologies to consider the changes taking place in Africa's food and agricultural production. It shows breeders how to use foresight to anticipate future demand requirements and incorporate these into new variety designs. Specifically, it provides a holistic approach to: (i) analysing the current agricultural landscape and challenges in Africa within a context of market supply and demand; (ii) understanding the drivers of change and their predictability; and (iii) using the methodology of social, technological, economic, environmental and policy drivers (STEEP analysis) and risk mitigation to create scenarios and validate new variety designs.

Visioning and foresight are important skills for a demand-led breeder. In order to create varieties that satisfy client and market needs, awareness of and competency is required in the use of the following:

- **Information sources.** Seeking and recognizing reliable information sources by investigating national statistics, regional documentation and reports, and data from international organizations that deal with food security and policy.
- **STEEP analysis.** Identifying and understanding social, technological, economic, environmental and political drivers of agricultural change.
- **Scenario creation.** Identifying unpredictable drivers and constructing scenarios of what the future agricultural landscape might look like based on these splitting factors.
- **Testing assumptions.** Testing the variety development assumptions against the scenarios created.
- **Risk mitigation.** Using risk mitigation and analysis as an essential procedure and a cornerstone of testing the sustainability of the forecasted requirements.

The chapter aims to enable plant breeders to: (i) understand the language and terms used in identifying drivers of agriculture; (ii) be discerning about the quality of data available; and (iii) critically analyse the implications of key conclusions from experts in the various disciplines that contribute to foresight analysis. It also aims to act as a resource for education in this field. For this purpose, boxes are included in several sections of the chapter that summarize their educational objectives and present the key messages and questions that are involved; in addition, there is a final box at the end of the chapter that summarizes its overall learning objectives.

Demand-led approaches to variety design depend heavily on the accuracy of the information obtained from participatory breeding approaches, stakeholders in the value chain and consumers. Typically, this market research information is provided within the context of current markets and consumer needs (see Chapter 3 this volume).

However, this chapter focuses more on future analyses by defining what the future demand and breeding technology landscape could look like. Both accessing detailed market research information and having a vision of what the needs and preferences of clients could be in the future are essential in new variety design. Current market data and future trends need to be treated iteratively to make decisions on breeding targets and new variety designs.

Further, this chapter focuses on broadening breeders' perspectives and increasing their awareness of current trends, opportunities and challenges for African agriculture, so as to put their work into a strategic context and see how it contributes to agricultural development and food security. This requires breeders to understand the current status and baseline of agriculture in their countries as the starting point for forecasting the future. These trends then need to be monitored and updated at least annually using the best available information sources and reliable data pertinent to the continent and to individual countries in Africa, so that breeding goals can be influenced continually.

## Agriculture in Africa: Outlook, Challenges and Policy (Box 2.1)

Agriculture in Africa faces many challenges. These have been summarized in recent reports (NEPAD, 2014, 2015) from the African Union's Comprehensive Africa Agriculture Development Programme (CAADP), which was endorsed by Africa's Heads of State and Government in 2003 (NEPAD, 2013). The African Union comprises all of the countries of the continent of Africa, and it provides Africa's policy framework for agricultural transformation, wealth creation, food security and nutrition, economic growth and prosperity for all; its technical body is NEPAD, the New Partnership for Africa's Development. The policy framework is a set of principles and strategies to help countries review their own situations and identify investment opportunities that have the optimal impact and returns. CAADP provides an evidence-based planning process with knowledge as a key primary input, and human resources development and partnerships as central factors. It aims to align diverse stakeholder interests around the design of integrated programmes adapted at the local level. Although it is continental in scope, CAADP aims to integrate national, regional and continental efforts to promote growth of the agricultural sector and its economic development.

The challenges highlighted in the CAADP strategy for 2015–2025 are:

- the food and nutrition requirements of African populations;
- economic inequality and poverty in rural areas;
- high population growth;
- the impacts of globalization on African agriculture, including the effects of climate change, markets and the search for new resources of green energy; and
- the maintenance of control over national natural resources.

### Strategies for transforming African agriculture

Several strategies are being advocated for transforming agriculture in Africa through national, regional and international fora and agencies that are active throughout the continent (FARA, 2014). These strategies include various combinations of:

**Box 2.1.** Agriculture in Africa – outlook, challenges and policy: educational objectives.

**Purpose:** to understand the concepts of market supply and demand; the outlook and challenges facing African agriculture; and how the business of plant breeding can contribute solutions to these challenges.

**Educational objectives:**

- to understand the concept and definition of supply and demand;
- to know the current status of African agriculture and the key challenges being faced in delivering agricultural development and food security;
- to be aware of the current status of supply and demand of key agricultural commodities in Africa and be able to access reputable data sources;
- to be familiar with national, regional and pan-African government policies to address food security and agricultural development and their respective targets;
- to understand targeted national R&D objectives and investment plans, and how plant breeding and improving crop genetics fits within this agenda;
- to decide how targeted plant breeding programmes can contribute to national goals; and
- to be able to access key data sources and monitor trends, issues and challenges in agriculture in targeted countries and across Africa.

**Key messages**

- The key message for breeders is to be aware of the changes taking place in African agriculture, especially in their own and neighbouring countries.
- Plant breeders need to know the policy landscape in which their work is situated and how crop improvement can contribute to food security and agricultural development.
- Breeders need to recognize that their entrepreneurialism can contribute to economic growth, improved smallholder farmer livelihoods and increased food security, i.e. that 'Improved varieties can change lives'.
- It is also important for breeders to know how to access reputable data sources and to have a baseline of information for the targeted country and investigate key trends. This information is essential for the creation of future scenarios and also in developing the business case(s) to support future public and private investments in demand-led breeding (see Chapter 7, this volume).

**Key questions**

- What is the current status of African agriculture?
- What are the key challenges facing African agriculture?
- What is the current balance of supply and demand for major African food crops?
- How do you measure supply and demand?
- What are the current public agricultural R&D policies at the national, subregional and continental level?
- What is the '*Science Agenda for the Transformation of Agriculture in Africa*' and how do plant breeding and demand-led approaches feature within it?
- How can plant breeding make a difference to lives and livelihoods in Africa?

- improving agricultural productivity;
- ensuring the availability and widespread use of quality farm inputs, including seeds;
- affordable access to new technologies, including biotechnologies;
- facilitating the growth of agricultural markets and trade;

- investing in public infrastructure for agricultural growth;
- reducing rural vulnerability and insecurity;
- improving agricultural policies and institutions;
- foresight and visioning to meet market/consumers' demands; and
- increasing investments in the agricultural sector, including through public–private partnerships.

The Forum for Agricultural Research in Africa (FARA) oversaw the development of the *Science Agenda for Agriculture in Africa* (FARA, 2014), a process that identifies the challenges and opportunities that science and technology can bring to the transformation of agriculture in Africa.

Breeders can contribute solutions to the current challenges by developing new varieties that improve:

- **productivity** – creating more productive varieties that meet consumer needs and preferences;
- **nutrition** – improving the nutritional composition of new varieties;
- **seed supply** – supporting the development of new varieties with traits that are in high demand by farmers and other value chain participants, thus enabling the emergence of seed retailers and distributors;
- **value creation** – supporting variety designs to satisfy different market segments, thus enabling greater market value and pricing structures for new entrants and new business development (see also Chapter 3, this volume);
- **value chains** – catalysing dialogue among government officials, farmers and value chain clients and stakeholders to find solutions for dysfunctional value chains;
- **business investment** – providing information on new varieties and creating designs that are targeted at attracting food processing companies to invest in targeted countries;
- **climate change adaptation** – addressing climate change trends in variety designs and supporting the analysis of crop improvement for resilience and/or shifting agricultural production to other crop species in response to climate change; and
- **capacity building** – supporting the education and training of new breeders so that there is sufficient human capacity in both public sector and the growing private sector breeding programmes in Africa.

*Public–private partnerships in transforming agriculture in Africa*

There is a new resolve by African governments to improve agricultural productivity and increase engagement and cooperation with the private agribusiness sector. There is also international recognition of the potential that Africa holds for agriculture This is evidenced by the number of high-level commitments that have been made to support the transformation of Africa's agriculture:

- **CAADP.** The African Union and participating countries are implementing CAADP, which includes building relationships with the private sector and encouraging regional trade agreements, and regional harmonization on regulatory matters (e.g. in the seed sector).

- **African Development Bank (AfDB).** The AfDB is supporting two initiatives to support the transformation of African agriculture, one on Technologies for the Transformation of African Agriculture (TAAT) and a second termed the African Agricultural Research Program (AARP) (see AfDB, 2016; FARA, 2016).
- **A Green Revolution for Africa (AGRA).** AGRA's portfolio on agriculture includes substantial support for plant breeding and seed systems through the Program for Agricultural Seed Systems (PASS) and agro-dealership development in Africa (SourceWatch, 2012).
- **New Alliance for Food Security and Nutrition.** This alliance was launched in 2012 (as a G8 initiative) and is a shared commitment between African governments, G8 governments and private industry to increase investment in Africa where the conditions are right to expand Africa's potential for rapid and sustainable agricultural growth (The White House, 2012; New Alliance, 2017).

## Agricultural supply and demand in Africa

### *Definitions*

SUPPLY AND DEMAND. This is one of the fundamental concepts of agricultural economics and it is the backbone of a market demand-led economy. Unmet demand is a precondition for the emergence of new businesses and local enterprise in the agricultural and food sectors in Africa.

DEMAND. Demand refers to the quantity of a product or service desired by buyers. The quantity demanded is the amount of a product customers are willing to buy at a certain price. The relationship between price and quantity demanded is known as the 'demand relationship'.

SUPPLY. Supply represents how much the market can provide. The quantity supplied refers to the amount of goods producers are willing to provide when they receive a certain price. The correlation between price and how much of a good or service is supplied to the market is known as the 'supply relationship'. Price, if unimpeded market forces are prevailing, is strongly influenced by supply and demand (Hayes, 2015).

At a national level, food demand can be calculated by multiplying population by per capita consumption. Domestic supply can be calculated by multiplying the area of crops grown by the total yield. When demand is greater than supply, if financing is available and the country has trade arrangements, then food can be imported to meet the unmet demand. However, for many countries in Africa, there is a food supply deficit and hunger is the consequence. During famines and other critical food shortage periods, food aid may be provided by foreign governments and international agencies, either in kind or as a grant to purchase food.

### *Supply*

Africa has 60% of the world's land that is suitable for agricultural production. In sub-Saharan Africa (SSA), small-scale farmers account for 80% of agricultural production and 70% of these farmers are women. Farm size is typically under 0.5 ha. Future food security and economic development is highly dependent on increasing the productivity of African smallholders. Farming is characterized by low inputs and low yields, with the average yields of many crops being well below the global averages. Although productivity has been increasing in some countries, it is not keeping pace with population growth, and food imports are being used to fill the gap, as supply cannot keep pace with demand. Africa's total food deficit is projected to increase to 60 million tons and cost US$14 billion by 2020 (AGRA, 2013).

Over the past decade, increases in production have come mainly from increasing the area of land used for agriculture rather than from productivity growth. The reasons for low productivity gains are many and complex (see Figs 2.1 and 2.2). They include the constraints listed under the five headings below:

#### INFRASTRUCTURE

- hard constraints: roads and port capacity, irrigation systems and access to water; and
- soft constraints: tariff barriers.

#### AGRICULTURAL PRODUCTIVITY

- lack of defined land ownership rights;
- surplus agricultural land that allows expansion, although often at the cost of the environment; and
- lack of government investment and poorly developed extension services.

#### AGRICULTURE INPUT MARKETS

- low quality seed;
- lack of use of crop protection inputs;
- lack of distribution networks for seeds, fertilizers and agrochemicals;
- few commercial private sector companies operating in many countries; and
- lack of credit for smallholder farmers to purchase inputs.

#### AGRICULTURE OUTPUT MARKETS

- lack of connectivity and access of farmers to markets;
- underdeveloped crop value chains for retailers, processors and food companies; and
- lack of bargaining power of farmers on prices.

#### AGRICULTURAL R&D

- lack of sufficient government investment in R&D;
- lack of improved varieties for many of Africa's staple crops;
- lack of capacity in modern molecular science; and
- insufficient R&D scientists and a brain drain to industrialized countries.

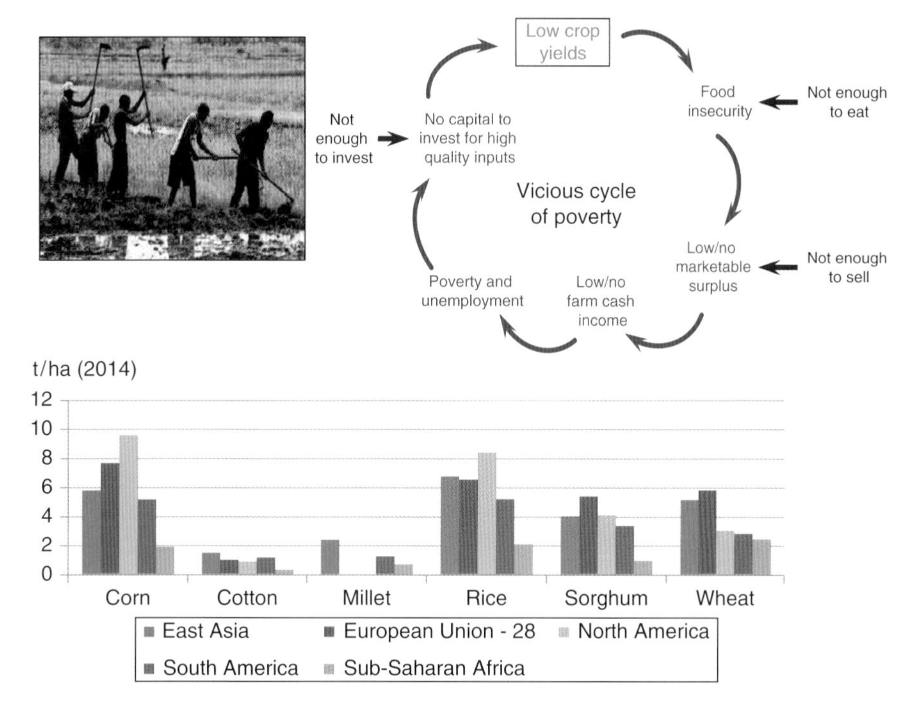


**Fig. 2.1.** Challenges of African agriculture: low productivity in smallholder farming. Data from PS&D (Production, Supply and Distribution), USDA Foreign Agricultural Service.

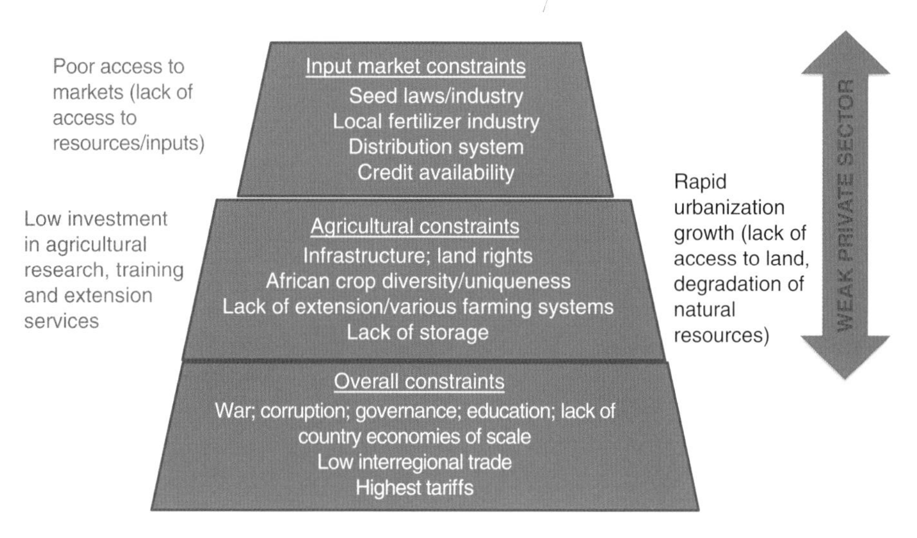


**Fig. 2.2.** Reasons for low productivity in African agriculture. Provided by J.L. Shoham.

### *Demand*

POPULATION GROWTH. Growth of demand in Africa is the fastest in the world and this, in turn, is driven by the highest rate of population growth. Population growth rates vary in different countries of Africa but the average is over 3% p.a. (see Fig. 2.3). Currently, there are about 800 million people in Africa and, of these, at least 220 million are undernourished. These rates are projected to continue and the population size is expected to double in 35 years. This compares with negative population growth rates in Europe and around 0.5% p.a. growth in North America, Asia and Latin America. A key feature of African population growth rates is that the growth in urban areas and cities is over double that in rural areas. This is of critical importance to plant breeders, as the design of new varieties must cater for the preferences of food industry suppliers and consumers who live in cities and towns. Demand-led breeding approaches can meet this need.

ECONOMIC GROWTH. The economies of African countries are growing. A primary indicator of economic growth is gross domestic product (GDP). GDP represents the monetary value of all goods and services produced within a country within a specified time period (Investopedia, 2015). Typically, it is calculated on an annual basis and is used to compare the economies of different countries. Currently, many countries in SSA have GDP growth rates of 4–8% p.a., compared with less than 2% in Europe and North America (see Figs 2.4 and 2.5) (World Bank, 2015). The agricultural sector contributes, on

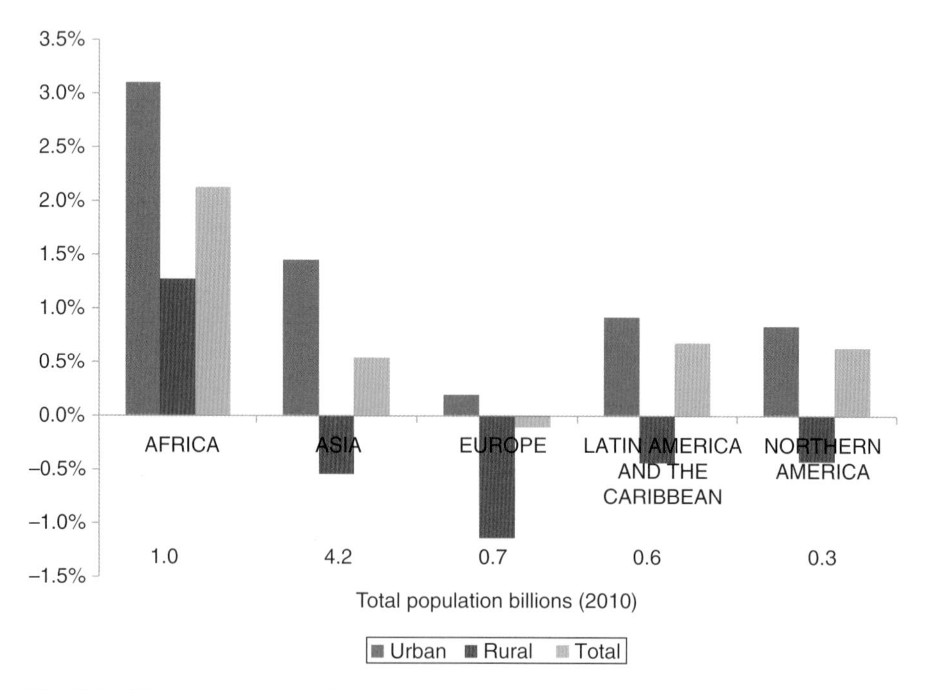


**Fig. 2.3.** Africa at a glance: the highest rate of population growth, as illustrated by the CAGR (compound annual growth rate) for 2010–2050. Data from the United Nations Population Fund (UNFPA).

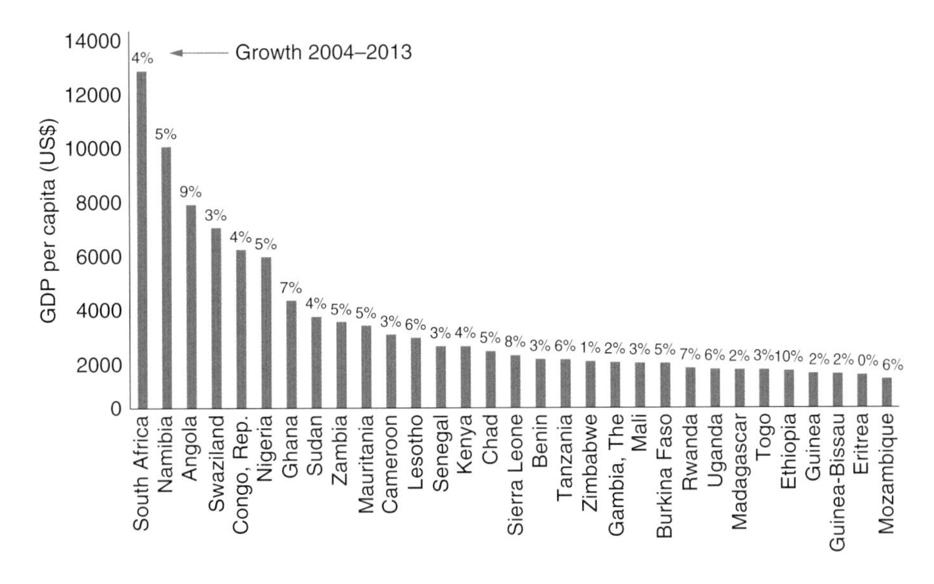


**Fig. 2.4.** Africa at a glance: high gross domestic product (GDP) per capita growth 2004–2013. Data from the World Bank, 2015; SFSA (Syngenta Foundation for Sustainable Agriculture) analysis.

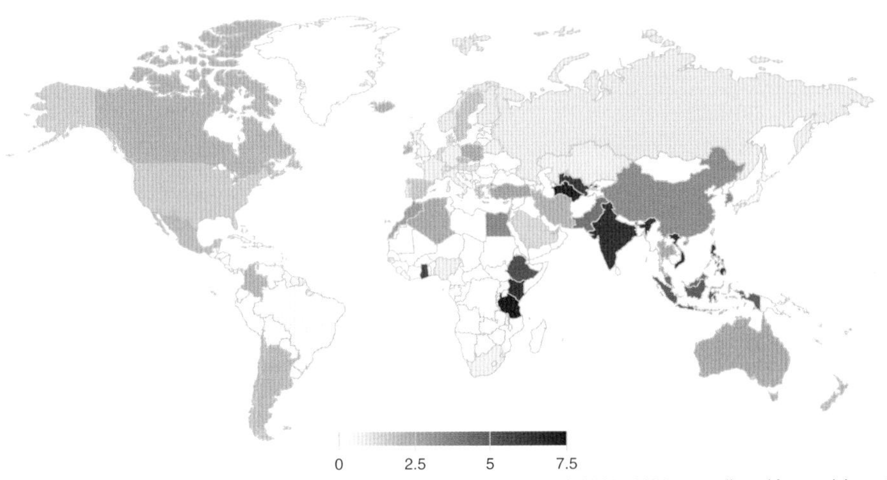


**Fig. 2.5.** Global outlook for gross domestic product (GDP) growth rates, 2016–2026. From The Conference Board Global Economic Outlook 2017.

average, 30–40% of total GDP and employs 60% of the total workforce, across the countries of SSA (FARA, 2014).

With economic growth and more money available for people to purchase food, consumer preferences change to higher demand for animal-sourced foods (meat, milk and eggs), as well as for more fruit and vegetables and convenience foods, especially among urban consumers. Producers are also more able to respond to the demands of the export market (Fig. 2.6).

## R&D capacity and investments in plant breeding in Africa

Africa has a unique, crop-based diet, with strong local, national and regional food preferences. Consumers in each country have their own preferences for staple crops and food recipes. Africa also has its own unique agro-ecological zones and a spectrum of specific pests and diseases that have required dedicated attention. Many crop varieties developed for other parts of the world often have limited utility in African ecosystems. However, access to the world's best germplasm and quality screening for specific characteristics and selection for utility is a good starting place for breeders working on food security crops in Africa.

Private sector seed companies invest little in R&D in plant breeding for many of these staple crops, especially where open-pollination systems, farmer-saved seed and clonal crops are the norm, and sustainable businesses and remuneration models involving hybrid seeds or technology fees for varieties with traits introduced through novel genetic modification (GM) are not feasible. The private sector does not have breeding programmes on staple food crops in Africa, such as bananas, yams, sweet potatoes, cassava, millet, pumpkins, indigenous African vegetables and many other such orphan crops. In these cases, crop improvement is wholly dependent on investments by African governments and international development agencies in the plant breeding programmes of African national agricultural research systems and the international agricultural research centres. Yet plant breeding in SSA provides a good example of the

- **Uganda**: growing apples, displacing imports
- **Zambia**: increase of cotton production
- **Kenya**: flower exports surpassed coffee exports
- **Ethiopia**: beans and coffee from local cooperatives responding well to international markets
- **DRC Congo**: post-conflict areas relying on cavies for nutrition and growth

**Fig. 2.6.** Economic growth in Africa: responding to new market demands. Cover shown of the World Bank's *Awakening Africa's Sleeping Giant* (2009). From The World Bank.

volatility of funding for R&D within that region (Lynam, 2010), where it is essentially a public sector enterprise, with limited sources of funds as well as unpredictable recurring support to guarantee continuity and longevity. These factors have an integral negative impact on the overall integrity of the plant breeding system.

### *Current plant breeding capacity in Africa*

There are approximately 800 plant breeders in SSA. More than half of these breeders (ca. 500) work in national plant breeding programmes in just four countries (South Africa, Nigeria, Ethiopia and Kenya); another 100 breeders are staff of the international agricultural research centres and regional programmes across Africa. The remaining 200 breeders are spread over some 40 countries of SSA.

Regional capacity building programmes in plant breeding aim to strengthen plant breeding capacity in SSA. These include the formal postgraduate programmes offered by the West Africa Centre for Crop Improvement (WACCI) in West Africa, the African Centre for Crop Improvement (ACCI) in eastern and southern Africa and Makerere University in Uganda for eastern Africa, as well as the professional development and research opportunities provided through Biosciences eastern and central Africa (BecA) and the CGIAR centres at the BecA regional biosciences hub in Nairobi.

## Seed Systems

Seed distribution systems are vital for smallholder farmers to be able to access improved varieties and quality seed. Africa's seed systems are at various stages of development, with only South Africa having a fully privatized seed sector similar to that in industrialized countries. A number of countries in eastern and southern Africa have evolving and progressive seed systems. Local seed companies (small and medium-sized enterprises (SMEs) and multinational agribusiness companies are investing in operations in Kenya, Uganda, Zambia, Malawi and Zimbabwe, with seed supply being most advanced for maize and soybeans (Figs 2.7 and 2.8). Most countries in SSA are at an early stage of seed system development and this is the major limiting factor preventing new varieties from reaching many farmers.

Private sector seed companies will often seek to employ an experienced plant breeder or agronomist as a starting point to understanding local opportunities and crop variety requirements. Typically, only once commercial operations are proving successful and profitability is reached will companies invest in their own local breeding programmes. The companies then look to recruit successful breeders from the public sector.

In developing countries, as with seed system development, there is a life cycle of evolution from public to private investment. In Africa, there is currently minimal private sector breeding taking place. As seed systems grow and become profitable, it is likely that there will be an increase in demand for professional breeders in the private sector. It usually follows in the longer term that the appetite of governments to finance the private sector's product pipeline (and

| | DuPont Pioneer | Monsanto | Vilmorin | Seed company | Syngenta | Others |
|---|---|---|---|---|---|---|
| Southern | | | | | | |
| South Africa | ✓ | ✓ | ✓ | ✓ | ✓ | |
| Zambia | ✓ | ✓ | ✓ | ✓ | ✓ | |
| Zimbabwe | ✓ | ✓ | | | | |
| Malawi | ✓ | ✓ | | ✓ | | |
| Others | Lesotho, Botswana, Angola | | | Angola, Botswana, Swaziland | | Baddar, Angola |
| Eastern | | | | | | |
| Kenya | ✓ | ✓ | ✓ | ✓ | ✓ | |
| Tanzania | ✓ | | | ✓ | | |
| Uganda | ✓ | | | | | |
| Ethiopia | ✓ | | | ✓ | | |
| Mozambique | Setting up | | | ✓ | | |
| Others | | | | Rwanda | | |
| North | | | | | | |
| Morocco | | | ✓ | | | |
| Tunisia | ✓ | | | | | Baddar |
| Egypt | ✓ | | | | | |
| Algeria | | ✓ | | | | Baddar |
| Libya | ✓ | | | | | |
| West | | | | | | |
| Nigeria | Setting up | | | ✓ | | Baddar |
| Ghana | ✓ | | | ✓ | | |
| Senegal | | | | | | Baddar |
| Others | | Burkina Faso | | DRC | | Baddar: Benin/Burkina Faso/Cameroon/ Chad/ Côte d'Ivoire/ Guinea/Mali |
| # Countries | 15 | <10 | NA | 15 | NA | Baddar: 15 Bayer: 8 |

**Fig. 2.7.** Global seed companies investing in Africa. From Shoham (2014). With permission from Informa's AGROW.

thereby profit) – by using public funds to support commercially viable plant breeding – declines. In Europe, North America and Australia, there is virtually no public plant breeding on core food crops, with the public sector focusing on pre-breeding activities, while private seed companies develop finished varieties. It is likely that a similar trend will evolve in those African countries where the agricultural sector transforms successfully from primarily subsistence agriculture to agriculture as a profitable business enterprise for small-, medium- and large-scale farmers. However, it is also likely that the breeding of certain food security crops will remain in the public sector for the foreseeable future, especially for clonally propagated crops with limited commercial potential for developing a seed business (e.g. banana, cassava, sweet potatoes).

## Visioning and Foresight Using STEEP Analysis and Scenario Creation (Box 2.2)

### Forecasting the future landscape by analysing drivers of change

*Overview*

Creating product profiles for new crop varieties based on market research and on consultations with farmers and all other key stakeholders in crop value

| Company | Maize | Veget | Wheat | Rice | Cotton | Sorgh | Soyb. | Others |
|---|---|---|---|---|---|---|---|---|
| Agricol | ✓ | | | | | | | Sunflower |
| Afrisem | | ✓ | | | | | | |
| Agri-Seeds | ✓ | | | | | | | |
| AGPY | | | | | | ✓ | | |
| Arab Sudanese Seed Company | | | | | ✓ | ✓ | | |
| Arda Seeds | ✓ | | | | | | ✓ | Cowpeas; Millet; Groundnuts; Sunflower |
| Baddar | | ✓ | | | | | | |
| Capstone | ✓ | | | | | | | Pasture/forage |
| East African Seed Co. | | ✓ | | | | | | |
| Ethiopian Seed Enteprise | | | ✓ | | | | | |
| Fica | ✓ | ✓ | | ✓ | | ✓ | ✓ | Millet; Groundnuts; Pasture |
| FreshCo | ✓ | ✓ | | | | | | |
| Funwe Seeds | ✓ | | | | | | ✓ | Pigeon peas; Cowpeas; Beans |
| Green Lakes Co. | | ✓ (toms) | | | | | | |
| Harvest Farm Seeds | ✓ | ✓ | | | | | | |
| Hygrotech | | ✓ | | | | | | |
| Kenya Highland Seed Co. | | ✓ | | | | | | |
| Kibo | | ✓ | | | | | | |
| Klein Karoo | | ✓ | | | | | | Pasture |
| KSC | ✓ | ✓ | | ✓ | | ✓ | | Millet; Pasture; Sunflower |

| Company | Maize | Veg. | Wheat | Rice | Cotton | Sorgh | Soyb. | Others |
|---|---|---|---|---|---|---|---|---|
| Leldet | | | | | | | | Groundnuts; Pigeon peas; Chickpeas |
| Link Seed | ✓ | | | | | | | |
| Maslaha Seeds | ✓ | | | | ✓ | ✓ | ✓ | Cowpeas |
| Monsanto | ✓ | | | | ✓ | ✓ | ✓ | |
| MRI | ✓ | | | | | | | |
| NASECO | ✓ | | | | | | | |
| Nectar Group | ✓ | ✓ | | | | ✓ | | |
| Otis Garden Seeds | ✓ | | | | | ✓ | | Millet |
| Premier Seeds | ✓ | | | ✓ | | ✓ | ✓ | Cowpeas; groundnuts; millet |
| Prime Seeds | ✓ | | | | | ✓ | | Millet |
| Pristine Seeds | ✓ | | | | | ✓ | | Cowpeas; millet; groundnuts |
| Reapers | | | | | | | | Groundnuts |
| Sakata | ✓ | | | | | | | Flowers |
| Seed Co. | ✓ | | ✓ | | ✓ | ✓ | ✓ | Cowpeas; groundnuts |
| Sesako | | | ✓ | | | | | |
| Syngenta | ✓ | ✓ | | | | | | Flowers |
| Technisem | | ✓ | | | | | | |
| Terratiga | ✓ | | | ✓ | | ✓ | | Cowpeas |
| Victoria Seeds | ✓ | ✓ | | ✓ | | ✓ | | Beans; millet; cowpeas; groundnut |
| Vilmorin | ✓ | ✓ | ✓ | | | | ✓ | |
| Western Seed Co. | ✓ | | | | | | | |
| Zamseeds | ✓ | ✓ | | | | | | |

**Fig. 2.8.** African seed companies and their crop portfolio. From Shoham (2014). With permission from Informa's AGROW.

**Box 2.2.** Visioning and foresight using STEEP analysis and scenario creation: educational objectives.

**Purpose:** to understand the core drivers that affect the supply and demand of food production and to create future agricultural scenarios using the STEEP (social, technological, economic, environmental and policy drivers) methodology.

**Educational objectives:**

- to define and describe drivers that may change the African agricultural landscape using the STEEP framework;
- to understand how these drivers influence supply and demand and can change future crop requirements, variety design and client needs and preferences;
- to identify the current and future commodity preferences for both producers and consumers;
- to identify key information sources and experts from across disciplines and assess the relevance and validity of information;
- to become aware of private sector seed companies and food companies operating in various countries in Africa and engage with them; and
- to create future scenarios and analyse their probability and impact on supply and demand.

**Key messages**

- Predicting the future is difficult and only in hindsight is there a 'right' answer.
- Using the STEEP methodology to build scenarios can help to avoid the creation of redundant varieties and assists confidence building for public and private investors, governments and R&D leaders.
- Scenario building is a way of forecasting possible futures. It is a methodology for exploring the full range of possibilities and the associated validity of plant breeding goals.
- Relying only on current market research information to predict the future is a high-risk approach.
- Identification and partnering with key experts in social trends, core breeding technology advances, economics, climate change, government policy and stakeholders in value chains is essential.
- Incorrect estimates or invalid assumptions about the future for each of the key drivers can invalidate the success of strategic plans.

**Key questions**

- What are the current drivers of agricultural change?
- What is the evidence supporting the existence of these drivers?
- What are the possible scenarios that could occur to change the demand for varieties?
- What are the demographic features or trends of the society (e.g. age distribution, lifestyle, education level and other social considerations)?
- How will climate change affect other drivers of agricultural change?
- What is the future for acceptance of GM biotechnology traits and new technologies in various countries?
- Will regional harmonization of seed registration requirements occur and, if so, when?
- How connected should breeders be with private sector seed companies and value chain participants?
- What will be the food deficit in Africa by 2025 if actions are not taken?

chains is the essence of demand-led plant breeding. The accuracy and reliability of market research and the use of farmer-participatory breeding methodology is linked to the time horizons of the respondents. In R&D terms, this is typically short for both farmers and seed-producing organizations. Farmers focus on their current activities and the problems they are facing during the current season and, potentially, during the next season. Seed producers may have a similar short time horizon, focusing on challenges over the next 1–3 years. For new variety releases, seed producers may have a 3–5 year outlook. However, the time frame for *de novo* breeding takes much longer. For most crops, and in most countries, new, demand-led plant variety designs developed by breeders today will need to remain relevant for at least 10 or more years into the future.

Some key questions for breeders participating in learning about demand-led variety development are:

- How relevant are the answers that you have been given by stakeholders in your crop value chain?
- Which design elements and drivers of change will remain stable over the time it will take to develop a new variety and which could change?
- How will you know if changes are taking place?
- How can you predict whether your designs will match the demands of future clients?
- How can you create a vision for what is needed in the future that has granularity and can be continuously monitored?

*The scenario creation approach*

Predicting the future is extremely difficult. However, creating different future scenarios by analysing drivers and the uncertainty linked to these drivers can prove to be extremely useful in determining the relevance and risks linked to new variety designs.

The driver analysis and scenario approach was developed by the Shell Company in the 1970s. It involves systematically identifying the full set of drivers that could change the future, and then, constructing potential alternative scenarios of how the future landscape could look. Organizations can then evaluate their product development strategies and new designs to see whether or not these will still be relevant to the market on the timescale required for new product creation.

The future will probably turn out differently from any of the scenarios created. However, the process of building the scenarios will: (i) increase awareness of the external factors that can affect the successful uptake of new varieties; and (ii) provide breeders with the possibility of testing and validating the applicability of their variety designs and development strategy against potential different agricultural future landscapes and their associated risks.

The process of forecasting and visioning the future by creating scenarios follows four steps:

**Step 1: Driver identification.** Identify the drivers of future agricultural change and organize them into five categories, namely social, technological, economic, environmental and political (STEEP) drivers, assigning a level of predictability to each one.

**Step 2: Information source identification.** Seek and evaluate the reliability of critical information sources.
**Step 3: Splitting factors and scenario creation.** Focus on the drivers that are unpredictable and are therefore considered to be 'splitting factors' to create scenarios of what the future agricultural landscape might look like. Regularly track and monitor splitting factors during the course of the breeding programme for further changes.
**Step 4: Variety specification validation.** Test new variety development assumptions against the scenarios created. These scenarios are then used as an input for decisions on the goals of the breeding programme and the design of the varieties.

STEP 1: DRIVER IDENTIFICATION AND PREDICTABILITY. Identify the drivers of future change for the targeted crop and country and organize them, using the following STEEP framework:

- social drivers – population growth, urban versus rural population trends;
- technology drivers – changes in the breeding scheme (e.g. type of parental lines, hybridization, genotyping and phenotyping technologies) and changes in the breeding approach (e.g. pre-breeding, core breeding and post-breeding evaluation);
- economic drivers – economic growth/GDP per capita, market sizes and value, food industry development, supermarket and retailer development, the emergence of seed distribution systems, crop price trends;
- environmental drivers – climate change, crop certification schemes; and
- political drivers – food self-sufficiency targets, government investment, seed harmonization regulation, food safety regulation, trade and exports, biosafety regulation for the use of GM crops.

*Social drivers.* Social drivers include population growth and urbanization, both of which have impacts on diets. The predicted rapid rise in population growth in cities and the changing food preferences of city and town dwellers are major drivers for the future. Few urban consumers will be growing their own food. Breeders need to consider the food preferences of urban dwellers when creating new variety designs and how they obtain this consumer information; in addition, they need to consider information on the agronomic and production requirements of the designs that is obtained through farmer participatory breeding programmes. The consumer acceptability of new products arising from the application of new technologies, including genetic modification to produce GM crops, is also a social driver that differs between countries and over time. Some examples of social drivers are given in Fig. 2.9.

*Technological drivers.* Technological drivers are those with which breeders are very familiar. The genomics revolution is creating many new possibilities to factor into future scenario planning. Often, both these technological drivers and the acceptance and use of new technologies are far from predictable.

| Driver | Impact | Predictability | Source |
|---|---|---|---|
| Population growth | Total demand | High | UN data |
| Urbanization | Dietary habits and tastes | High | UN data |
| GM acceptability and regulation | Technical possibilities | Low | IFPRI, ISAAA, news media |

**Fig. 2.9.** Social drivers of change for new crop varieties. Key: GM, genetic modification; IFPRI, International Food Policy Research Institute; ISAAA, International Service for the Acquisition of Agri-biotech Applications; UN, United Nations.

These drivers factor in all changes over the overall breeding methodology/cycle that have an impact on the rate of genetic gain. The genetic gain is the amount of increase in performance. It is improved by a sole increase or by a combined increase of the selection intensity, selection accuracy and genetic variance, and/or by a reduction in the number of years per breeding cycle. These drivers include changes in pre-breeding, core breeding and post-breeding evaluation. Pre-breeding involves genetic diversity, trait identification and marker development. Core breeding relates to: (i) the development of either fixed lines and hybrid lines (the selection of existing natural parental lines or genetic variants, recombination of genes/genomes and selection of best recombinants through accurate phenotyping and high throughput genotyping/DNA marker technology) or (ii) population breeding, including shuttle breeding. Post-breeding evaluation deals with selecting 'mega' varieties and selected lines per target population environments (TPEs).

Approaches that will shape the evolution of classical breeding into speed/smart breeding for the future include the use of:

- **New molecular markers**, such as simple sequence repeat (SSR), expressed sequence tagged (EST) and single nucleotide polymorphism (SNP)-based markers, which will increase selection accuracy.
- **High throughput**, low-cost sequencing, genotyping and phenotyping technologies to deal with large populations, which will reduce phenotyping error and therefore increase the intensity and accuracy of selection.
- **An informatics pipeline**, which will provide a meaningful sense of the huge amount of genomic data generated through the sequencing platforms after data curation and analysis.
- **Genome-wide selection** (GWS) and **genomic selection** (GS) – two prediction-based breeding methods that are used, respectively, to develop

genome-wide population maps, and to produce more accurate typing than plain phenotyping, which will reduce the number of years per breeding cycle.

  - GWS is a marker trait association technique using historical phenotypic data on a given population. It measures and analyses DNA sequence variations across a target subpopulation genome from a given population to identify traits that have common association in this population. The ultimate goal of GWS is to use this linkage to make predictions about which individuals in the population may have the trait of interest for the development of new varieties.
  - GS is an innovative plant breeding method that uses statistical modelling to predict how a plant will perform before it is field tested. The performance of the given line is estimated based on its genomic estimated breeding values (GEBVs), which depend on the accuracy of the statistical model used for its computation.

- **Data integration and decision support platforms**, which include the Integrated Breeding Platform (IBP), the CGIAR Excellence in Breeding (EiB) platform and the Integrated Genotyping Services and Support (IGSS), and provide access to the tools and support needed by breeders.
  - The IBP is a comprehensive suite of tools (the Breeding Management System, BMS) and services (e.g. genotyping) developed by the CGIAR Generation Challenge Programme (GCP) and its partners to help breeders in their day-to-day activities to manage breeding activities efficiently from project planning to statistical analysis and decision support via data collection, storage, curation and management. The stand-alone version of the BMS is already used by over 200 breeders in developing countries.
  - The new EiB, led by the International Maize and Wheat Improvement Center (CIMMYT), aims to bring high-throughput genotyping, phenotyping and other services, to large breeding programmes in the CGIAR system.
  - The IGSS is a joint venture between the BecA platform at the International Livestock Research Institute (ILRI) in Nairobi and the Diversity Array Technology (DArT, from the Australian company of that name) for the development and delivery of breeding support services based on a cost-efficient and affordable business model. The platform also integrates related informatics data management, data analyses and decision-support services for the investigation of common/orphan crops and complex genomes. It also provides a wide range of applications from simple gene/quantitative trait locus (QTL) fine mapping to complex genomic selection via genomic prediction and genome-wide association study.

Some examples of technological drivers are given in Fig. 2.10.

*Economic drivers.* Economic drivers cover factors such as crop prices, economic growth as measured by GDP per capita (GDP), and investments in the food chain – from the seed sector through to investments in food processing and food retail. These factors influence the demand side for crops and the

| Driver | Impact | Predictability |
|---|---|---|
| Biotechnology | Genetic variance, speed and cost | Low |
| High-throughput phenotyping | Selection intensity, number of years per breeding cycle | Low |
| High-throughput genotyping | Selection accuracy, breeding speed and cost | High |
| Pre-breeding | Breeding possibilities | Low |
| Core breeding | Breeding possibilities | Low |
| Post-breeding | Breeding possibilities | Low |
| Big informatics data | Data management and analysis | Low |

**Fig. 2.10.** Technological drivers of change for new crop varieties.

supply side for seeds. Some of these drivers, such as the GDP-related ones, are predictable; others, such as the development of infrastructure and the agro-dealer network are less predictable. Some examples of economic drivers are shown in Fig. 2.11.

There is a strong correlation between GDP per capita and the types of food consumed. Rising incomes typically means increased demand for and consumption of animal-sourced foods (meat, milk and eggs) and fruits and vegetables in diets. GDP growth in Africa is expected to remain high in the future, but this needs to be considered on a country-by-country basis. As the food processing and retail sectors develop, they also create new crop requirements. The development of the food processing and retail sectors in Africa reflects a general trend in increasing foreign direct investment on the continent (Fig. 2.12).

*Environmental drivers.* Environmental drivers relate to key parameters such as climate change, water availability and pest incidence (Fig. 2.13). All of these affect crop production and the supply side of the supply and demand equation. Despite the efforts being made to better understand and model climate change, the future changes remain largely unpredictable. Breeders need to connect with the experts in climate change who are working on forecasting models in their own country and use the emerging information in their breeding programmes. Specifically, focus is required on the timescales being predicted for change. If these are within the timescale of the breeding programme and delivery of the new variety, then assumptions on factors such as future drought resilience and temperature requirements should be built into the trait requirements in the variety design.

| Driver | Impact | Predictability | Source |
|---|---|---|---|
| GDP/capita | Food consumption patterns | High | World Bank; FAO Food Balance Sheets |
| Food industry /retailer development | Demand for improved seeds, AMCs<br>Scope for PPPs | Medium | Reardon (2011) |
| Seed company developments | Seed improvement | Medium | Informa (2014) |
| Dealer network | Accessibility of seeds | Low | AGRA (2013) |

**Fig. 2.11.** Economic drivers of change for new crop varieties. Key: AMCs, advanced market commitments; PPPs, public–private partnerships.

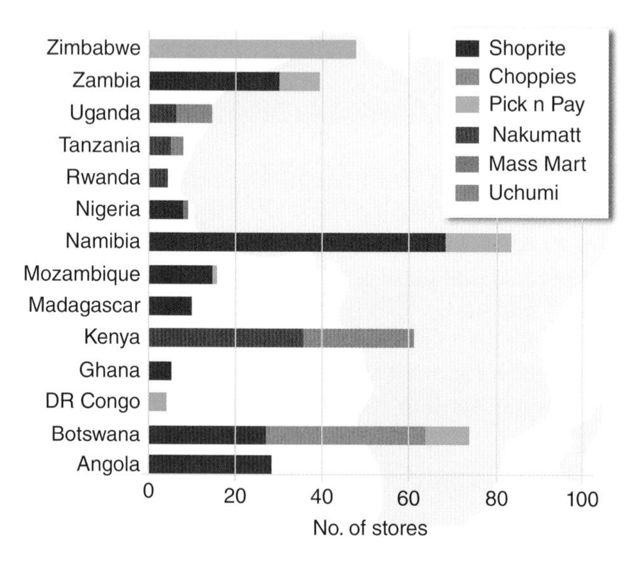


**Fig. 2.12.** Selected multi-country food retailers (supermarkets) in Africa. From *Promar Insight*, June 2014 (Promar International, 2014).

Crop and food certification schemes, which are proliferating, can also be included in this category of drivers. Both can have a major impact on crop production requirements and access to both domestic and international markets.

Some key questions concerning environmental drivers are:

- What are the predictions for climate change in the target country and what will be the impact?

| Driver | Impact | Predictability | Source/ Milestones |
|---|---|---|---|
| Climate change | Crop yields<br>Agronomic traits<br>Extreme events | Low | IPCC/Paris, 2015 |
| Certification schemes | Traceability<br>Food safety<br>Export market access | Medium | |
| Pest incidence | Crops yields and quality | Low | CABI Plantwise |

**Fig. 2.13.** Environmental drivers of change. Key: CABI Plantwise, a global programme led by CAB International that works to help farmers lose less of what they grow to plant health problems; IPCC/Paris 2015, United Nations Intergovernmental Panel on Climate Change/Paris Agreement within the United Nations Framework Convention on Climate Change.

- Will the cropping systems need to change?
- What are the specific environmental parameters that need to be factored into new variety design, e.g. water availability at a specific growth stage, heat or cold tolerance, light intensity, etc.?
- What food certification requirements will be needed for food processing companies or retailers for exports?

*Political drivers.* Political drivers embrace seed regulations, agricultural subsidies and tariffs, all of which have an impact on both supply and demand for crops, and trade-related developments such as regional trade agreements. Regional and national R&D investment strategies also are key drivers for breeders. Political drivers are inherently unpredictable. Some examples of political drivers of change are shown in Fig. 2.14.

*The application of STEEP drivers.* Different STEEP drivers will have an impact on different aspects of plant breeding. Plant breeders need to be aware of the key drivers influencing the demand and supply equation in the target country for their key crops and how this could affect their targets of choice and their crop designs. Breeders should also be tracking the changing array of technology options and solutions that could be pertinent to addressing their breeding goal(s). In terms of access to new scientific developments and new technologies, variety development, capacity strengthening and the delivery of new varieties to farmers, breeders need to decide on the preferred, most cost-effective approach, which may be through in-house solutions, contract outsourcing, partnering or a combination of all three of these approaches.

Having been through the above process of identifying the drivers and analysing their predictability, plant breeders should now be in a position to

| Driver | Impact | Predictability | Source |
| --- | --- | --- | --- |
| National seed laws | IP protection Private sector investments | Low | SeedQuest |
| Regional seed/variety harmonization schemes | Development costs, speed of variety release | Low | COMESA, ECA, ECOWAS SADC |
| Agricultural policies (CAADP) | Investment focus | Low | CAADP website |
| Nutrition policies | Consumer traits | Medium | IFPRI |

**Fig. 2.14.** Political drivers of change. Key: CAADP, Comprehensive Africa Agriculture Development Programme; COMESA, Common Market for Eastern and Southern Africa; ECA, Economic Commission for Africa; ECOWAS, Economic Community of West African States; IFPRI, International Food Policy Research Institute; SADC, Southern African Development Community; SeedQuest, a central information website for the global seed industry.

construct scenarios that could affect their own breeding programmes, to track ongoing developments for the most important drivers and to adjust their variety designs accordingly over time.

STEP 2: RELIABLE INFORMATION SOURCES. For each key driver, breeders need to seek input from experts in that discipline in their country so that they have the best information and expert advice to make decisions on their breeding goals, targets and variety specifications. Who and where are the experts in the country? Are they in the breeder's home institution or elsewhere? Breeders should enquire about the most reliable information sources that can be used for national statistics and other drivers of change, and would help to reach an understanding of the speed at which changes can occur. National agricultural investment plans and other CAADP documents should be consulted for information on national priorities as an important driver. When considering future technology and science drivers, it may be useful for breeders to consult with pan-African and international experts in the field; if so, who are they? Breeders also need to ask whether they are connected with the best and most successful breeders of the relevant crop worldwide who may be implementing new approaches, and whether there are approaches that can speed up the breeding programme if resources or new germplasm can be found.

Clearly, national statistics are the place to start. It is worth questioning how these data have been generated. The data may not always be accurate. For example, data from actual farming surveys are likely to be more robust than data obtained from expert panels. Data from the World Bank, the International Food Policy Research Institute (IFPRI), the United Nations (UN) Food and Agriculture Research Organization (FAO) and other UN agencies are reputable and well used. It is often best to consider ranges of values rather than absolutes and to understand the range of values where the assumptions used for decision making still hold true. Also, data can be interpreted in different ways. Reading only the summaries of reports can be misleading. Breeders do not need to be skilled in all disciplines related to drivers of change, but they do need to be able to understand the language and terms used and be discerning about the validity of data, and be able to think critically about the implications of key conclusions from experts in other disciplines.

STEP 3: SCENARIO CREATION AND THE USE OF SPLITTING FACTORS. A splitting factor is a driver that is difficult to predict. STEEP methodology uses splitting factors to create different scenarios of what the future can look like. An example of a splitting factor driver is consumer acceptance and government support for the commercialization of GM crops in African countries. The risks and benefits issues and regulatory requirements are understood by scientific experts. Consumer surveys and other market research will demonstrate the range of consumer views, but it is extremely difficult to know whether governments will provide registration approvals, on what time frames these may be provided and at what cost. Depending on the acceptance landscape of the target country, selecting a breeding strategy that is wholly dependent on the acceptance of GM crops in that country and finding an investment partner to finance the costs of meeting regulatory requirements, including any environmental studies, will depend on many assumptions. These assumptions may or may not be fulfilled and are difficult to predict. Therefore, as a core dependency for success, a breeding strategy wholly based on using GM technology may have high risk. However, the risk could be lowered by including GM and other new technologies along with more conventional breeding as a broader breeding strategy for developing a suite of new crop varieties with targeted traits.

### *Splitting factor 1: a regulatory and consumer acceptance driver*

Scenario 1:

- A regulatory framework for genetically modified organisms (GMOs) is in place or expected to be approved within 5 years.
- The targeted trait is not the first to reach the market in Africa.
- No other technology can provide the disease resistance (or another essential trait) required and smallholder farmers lobby the government for approvals to be given.
- The trait has been already registered and product consumed in five countries of the Organisation for Economic Co-operation and Development (OECD).

Scenario 2:

- It is unclear when the regulatory framework of the GMO will be approved. No regulatory approvals are likely in-country over the next 10 years.
- The targeted trait is the first to be tested and reach the market in Africa.
- Some improved performance may potentially be achieved by searching for resistance in wild relatives and incorporating this trait into the crop plant through wide crossing.
- The trait is registered and used in the USA but not in countries of the European Union (EU).

*Splitting factor 2: molecular breeding technology uptake and adoption by breeders*

Scenario 1:

- The facilities and skills for molecular breeding are in place or are expected to be established within a short time.
- Molecular breeding technologies are not the first new technologies to be applied and adopted by African scientists.
- No other technology can efficiently address the constraints (or several essential traits at once) and breeders are willing to apply the technology.
- The molecular breeding approach is used worldwide in advanced research institutions and has proven efficient and effective in modern breeding.

Scenario 2:

- It is unclear when facilities and skills will be built and provided to breeders. There are no indications of initiatives in progress to develop the required facilities and skills.
- Molecular breeding is in its infancy of application and adoption in most African breeding programmes.
- Other innovative breeding schemes can be investigated by increasing genetic gain by increasing the number of crop cycles a year and the intensity of selection using high-throughput phenotyping.
- Most steps of the cultivar development pipeline that use molecular breeding can be outsourced or implemented in partnership with advanced research institutions outside Africa.

For each of the above two types of splitting factor, a consideration of two different scenarios of the regulatory and consumer environment and/or new technology adoption, as well as the technical challenges, enables a decision to be made on which is the more likely scenario for the country/crop/trait concerned. This process will influence the decisions made on new variety design and the preferred technology approach.

*Signposts.* Each of the unpredictable drivers/splitting factors discussed above should be monitored by identifying signposts that can help breeders predict which way the outcome is heading. The creation of a network of experts and information sources that will be alert to changes taking place is advisable.

These networks can operate at a national, regional and/or global level. Some examples of signposts are:

- mutual recognition of variety registration data within a region such as the Common Market for Eastern and Southern Africa (COMESA);
- new seed laws in place;
- trade harmonization agreements; and
- the launch of a major GM food crop in one or more African countries (e.g. GM maize).

In addition to tracking selected signposts, it is important to incorporate review points in the development plans of the breeding programme to ensure that the lead breeder actively take decisions in response to external risk factors. Some scenarios will mean that the traits selected in a new variety design may need to be amended over time. Alternatively, the analysis may confirm that the combination of traits selected is the right one with which to proceed.

STEP 4: VARIETY SPECIFICATION VALIDATION. This step involves reviewing the assumptions used in the new variety design and testing these against each scenario. Specifically, this includes revisiting the timelines required to develop and release a new variety and the likely time frames for each forecast. In the case of *de novo* design, this includes deciding which traits will remain relevant for the most likely scenarios of the future, taken over 10–12 years.

## Integrating Foresight into New Variety Design (Box 2.3)

This chapter (Chapter 2) and the variety design tool described in Chapter 4 of this volume together provide the systematic framework for breeders to create new designs. Each aspect of the product profile should be reviewed using the time horizon it will take to create a new variety within the breeding programme.

First, in the light of the future scenarios created, the crop segment should be re-examined to see whether the client base, location and market are expected to remain as predicted, or whether significant changes are likely to occur.

Second, each trait characteristic should be examined in the light of future predictions. It may be that some traits change in terms of their relative value and importance within the breeding programme. Some traits may become redundant, while other traits may need to be included in the profile. The quantitative benchmark for success for each trait may also need to be adjusted in the light of new information. The practical output from the forecasting and visioning exercise is a revised product profile for each client segment that is fit for purpose, with an associated monitoring plan to track changes in requirements and milestones for review within the breeding stage plan over time (see also Chapter 5, this volume).

**Box 2.3.** Integrating foresight into new variety design: educational objectives.

**Purpose:** to ensure that the design of all new varieties that are developed has taken into account potential changes in demand within their agricultural setting over the next decade.

**Educational objective:** to integrate long-term drivers into the creation of new variety designs and priority setting within breeding programmes.

**Key messages**

- Foresight methods can be used to validate or review existing variety designs or as a starting point for the creation of new designs. Both approaches are equally valid.
- Clarity is needed on who the variety is being designed for and whether this client group will change.
- A detailed review is required of every trait within an individual variety design, and the product profile should be amended accordingly.

**Key questions**

- Who will be the clients and what will their needs be in 5–10 years?
- Will a new variety designed today still serve the needs of the farmers and value chains that it was designed for, given the time it will take to be developed, registered and released?
- Which traits should be changed or included in the product profile based on different future scenarios?

## Managing Risks (Box 2.4)

### Types of risk

The main risks associated with forecasting approaches come from lack of demand and failure of supply.

*Lack of demand*

- Incorrect assumptions are made about market demand, based on inaccurate or inadequate information.
- The critical number of clients with demand decreases or preferences change.
- Easier solutions are found to address clients' needs instead of using genetics.
- Better varieties are produced by somebody else that meet clients' needs and preferences.

*Supply failure*

- Technology promises fail to deliver.
- The development time takes much longer or more resources are required than was expected.
- Farmers and consumers' preferences have been inaccurately identified.
- The cost of the product to consumers is unaffordable.

**Box 2.4.** Risk management: educational objectives.

**Purpose:** to understand the importance of risk management in new variety design and also the importance of taking risk mitigation actions to ensure that demand for the new varieties will endure during the variety development process.

**Educational objectives:**

- to identify the range of risks and uncertainties in a variety design when using STEEP (social, technological, economic, environmental and policy) driver analysis and scenario-based forecasting;
- to evaluate and prioritize these risks; and
- to decide on preventive actions to avoid or mitigate these risks.

**Key messages**

- Review key success criteria and the associated risks in targeted variety designs. Look at every trait characteristic in the design and decide whether it will remain relevant for its intended users during the time required for variety development.
- Understand the benefits and costs of maintaining many biologically diverse germplasm lines to spread the risk. Integrate decision points into the stage plan (see Chapter 5, this volume).
- Risk analysis and mitigation is an essential component for testing the longer term viability of variety designs.

**Key questions**

- What are the main potential risks in the targeted variety design?
- What are the associated potential impacts and consequences of these risks?
- Who and what will be affected by these consequences?
- How can the breeder manage or mitigate these risks?

## Risk mitigation options

*Transparency of assumptions*

Clarity is essential in the core assumptions made about the vision of the future agricultural landscape and the market environments into which a new variety will enter. There should be a list of specific assumptions available on the range of success criteria that are being used to design the target variety, including:

- the variety development timetable;
- its date of registration and availability to farmers;
- the numbers of farmers expected to have access to the variety and who will want to use it;
- the emergence of seed producers and distributors;
- the entry of private sector versus public distribution systems;
- the availability of alternative varieties;
- the acceptance of the specific technology used by consumers; and
- access to technology and trait germplasm sources.

*Monitoring and early warning systems*
A monitoring system is essential for each of the key drivers and risk assumptions, in order to provide early warning if the assumption is proving to be incorrect. Continuous interaction is required with value chain participants and government officials who influence the enabling environment, in areas such as trade policy and regulatory systems. Issues like changing trade policies and food safety regulation take time for governments to achieve. By monitoring the enabling environment and its specific elements, emerging risks should be evident well in advance and mitigation measures put in place.

*Risk mitigation options*
Given future uncertainties, there are a number of actions that can be considered for the mitigation of risks, such as:

- maintaining a diverse set of variety selections that meet the broadest set of criteria for the scenarios created, and the areas of uncertainty, for as long as is economically possible in the breeding programme;
- maintaining awareness by key clients in the value chain of the lead germplasm selections and involving them in the progressive decision making on the selection of lines going forward;
- regularly reviewing and testing the key assumptions underlying variety selection; and
- encouraging investment that supports the diversification of breeding technology tools/platforms that are additional to conventional breeding techniques.

## Learning Methods (Box 2.5)

Before this chapter concludes, a summary is provided in Box 2.5 of learning methods – together with assignments and assessment methods – for each of the main topics that have been covered in the main sections of the chapter: Agriculture in Africa: Outlook, Challenges and Policy; Visioning and Foresight Using STEEP Analysis and Scenario Creation; Integrating Foresight into New Variety Design; and Managing Risks.

**Box 2.5.** Learning methods, assignments and assessment methods.

**Agriculture in Africa: Outlook, Challenges and Policy**

*Learning method*

- PowerPoint presentation on principles of supply and demand and challenges facing African agriculture.
- Web search of literature and related information for group discussions on supply and demand for staple food and cash crops.

*Continued*

**Box 2.5.** Continued.

*Case studies*

- Search and review relevant case studies on demand and supply on relevant crops in or outside Africa.

*Assignment*

- Review the CAADP (Comprehensive Africa Agriculture Development Programme) strategic implementation plan (2015–2025) and national investment plans and highlight how a national plant breeding programme can contribute to the delivery of targets.

*Assessment*

- Assignment.
- Exam questions on the key principles of supply and demand and the challenges facing African agriculture.

**Visioning and Forecasting Using STEEP Analysis and Scenario Creation**

*Learning method*

- Power point presentation to learn the STEEP (social, technological, economic, environmental and policy) process.
- Group learning by using the STEEP process to identify the main drivers of change that could affect the group's crop breeding programmes and whether the drivers affect the supply or demand side of their variety design assumptions. The creation of scenarios and identification of splitting factors for monitoring.
- Web search for literature on drivers and splitting factors.
- Selected reference documents (see list at end of this chapter) may be used to discuss a range of types of drivers that can influence the design of new varieties.

*Assignment*

- Participants in the learning process to investigate the relevant literature and speak to experts in the field to identify the most important drivers of change and create a set of scenarios of the agricultural landscape that are relevant for their crop breeding programme in their country. Also to suggest probabilities for each driver and highlight the drivers and splitting factors that they will monitor as professional breeders.

*Assessment*

- *Ex ante* evaluation on the breadth of knowledge of participants on agricultural drivers
- *Ex post* evaluation via exam questions on the core principles of: (i) supply and demand; (ii) drivers of agricultural change; (iii) the STEEP method and use of splitting factors to create future agricultural scenarios that can affect plant breeding programmes.

**Integrating Foresight into New Variety Design**

*Learning method*

- Group discussions on integrating longer term requirements into variety designs and plant breeding programmes.
- PowerPoint presentation giving examples on how to do this.

*Continued*

**Box 2.5.** Continued.

*Assignment*

- Participants to review the specific trait characteristics in their designs and highlight those that have the greatest uncertainty. To search for relevant expert inputs and information from the literature and the Internet and decide whether changes should be made to the design(s). Also, to review whether crop breeding is the most appropriate way forward or whether there are easier and quicker solutions that can be found as viable alternatives to meet clients' needs.

*Assessment*

- Assignments.
- Exam questions on how to incorporate foresight into product profiles/new variety design, using examples to test the decision making of breeders.

**Managing Risk**

*Learning method*

- PowerPoint presentation on core principles.
- Web search of literature and related information on risks associated with visioning and forecasting methods.
- Group discussions to explore risks and mitigation methods using assumption-based supply and demand approaches.

*Case studies*

- Ghana tomato processing case study (Baba *et al.*, 2013).
- Seek case studies for learning from participants attending demand-led continuing professional development workshops (see assignment below).

*Assignment*

- Participants to seek examples of varieties registered at their home institutions that have a low track record of adoption by farmers. To identify what were the primary reasons for low adoption and rank these by importance. To classify the reasons into client demand or seed supply reasons. When the primary reason is a demand failure, then to investigate in detail if/what design features reduced the interest of farmers or the supply chain. To share these examples for future professional development courses.
- Practical session for participants to identify and evaluate potential risks facing their own breeding programmes, highlight specific assumptions being used in their variety designs and outline the actions and risk reduction measures that should be implemented in the programme.

*Assessment*

- Assignment
- Exam questions on the importance of risk mitigation measures and maintaining early warning systems that detect changes in the assumptions used in new varieties under development.

## Resource Materials

Slide sets are available for this chapter as part of Appendix 3 of the open-resource e-learning material for the volume. These summarize the chapter contents and provide further information. The e-learning material is available at http://www.cabi.org/openresources/93814 and also on a USB stick that is included with this volume.

## References

AfDB (2016) "Africa Feeding Africa": New mega initiative to transform agriculture in Africa is focus of international meeting. African Development Bank Group, Abidjan, Côte d'Ivoire. Available at: https://www.afdb.org/en/news-and-events/africa-feeding-africa-new-mega-initiative-to-transform-agriculture-in-africa-is-focus-of-international-meeting-15583/ (accessed 1 May 2017).

AGRA (2013) *Africa Agriculture Status Report 2013: Focus on Staple Crops*. Alliance for a Green Revolution in Africa (AGRA), Nairobi.

AGRA (undated) AGRA's Program for Africa's Seed Systems (PASS): Strengthening Public Crop Genetic Improvement and Private Input Supply Across Africa. Alliance for a Green Revolution in Africa (AGRA), Nairobi. Available at: http://www.fao.org/fileadmin/user_upload/drought/docs/AGRA%20Seed%20Systems%20and%20the%20future%20of%20farming.pdf (accessed 2 May 2017).

Baba, I.I.Y., Yirzagla, J. and Mawunya, M. (2013) The tomato industry in Ghana: fundamental challenges, surmounting strategies and perspectives – a review. *International Journal of Current Research* 5, 4102–4107.

Economist (2016) *The World in 2015*. The Economist Group, London.

FARA (2014) *Science Agenda for Agriculture in Africa (S3A): "Connecting Science" to Transform Agriculture in Africa*. Forum for Agricultural Research in Africa (FARA), Accra, Ghana. Available at: http://faraafrica.org/wp-content/uploads/2015/04/English_Science_agenda_for_agr_in_Africa.pdf (accessed 2 May 2017).

FARA (2016) Proposal Development Workshop of the African Agricultural Research Programme (AARP). Forum for Agricultural Research in Africa (FARA), Accra, Ghana. Available at: http://faraafrica.org/news-events/proposal-development-workshop-of-the-african-agricultural-research-programme-aarp/ (accessed 1 May 2017).

Haynes, A. (2015) Economics Basics: Supply and Demand. Available at: http://www.investopedia.com/university/economics/economics3.asp (accessed 1 May 2017).

Investopedia (2015) Gross Domestic Product – GDP. Available at: http://www.investopedia.com/terms/g/gdp.asp (accessed 1 May 2017).

Lynam, J. (2010) *Evolving a Plant Breeding and Seed System in Sub-Saharan Africa in an Era of Donor Dependence. A Report for the Global Partnership Initiative for Plant Breeding Capacity Initiative*. FAO Plant Production and Protection Paper 210. Available at: http://www.fao.org/3/a-at535e.pdf (html version accessed 16 May 2017).

NEPAD (2013) *Introducing the Comprehensive Africa Agriculture Development Programme (CAADP)*. New Partnership for Africa's Development, Midrand, South Africa. Available at: http://www.nepad.org/download/file/fid/3606%20 (accessed 2 May 2017).

NEPAD (2014) *Sustaining CAADP Momentum – Going for Results and Impact. The CAADP 10-Year Results Framework*. New Partnership for Africa's Development, Midrand, South Africa. Available at: http://www.merid.org/en/Africanagricultureandfoodsystems/~/media/Files/Projects/Africa%20Ag%20and%20Food%20Systems/Sustaining%20CAADP%20Momentum%20Results%20Framework%20-%20Feb%202014%20-%20consultation%20Version.pdf (accessed 2 May 2017).

NEPAD (2015) *Implementation Strategy and Road Map to Achieve the 2025 Vision on CAADP*. New Partnership for Africa's Development, Midrand, South Africa. Available at: http://www.nepad.org/download/file/fid/4282 (accessed 2 May 2017).

New Alliance (2017) New Alliance for Food Security and Nutrition. Available at: https://new-alliance.org/ (accessed 1 May 2017).

Promar International (2014) *Promar Insight* No. 6, June 2014, p. 4. Available at: http://www.promar-international.com/_userfiles/publications/files/Promar Insight Africa (accessed 1 May 2017).

Reardon, T. (2011) The global rise and impact of supermarkets: an international perspective. Paper prepared for presentation at: *The Supermarket Revolution in Food: Good, Bad or Ugly for the World's Farmers, Consumers and Retailers? Conference conducted by the Crawford Fund for International Agricultural Research, Parliament House, Canberra, Australia, 14–16 August 2011. The Crawford Fund 2011 Annual Parliamentary Conference.* The Crawford Fund, Canberra, Australia. Available at: http://ageconsearch.umn.edu/bitstream/125312/1/Reardon2011.pdf (accessed 1 May 2017).

Shoham, J.L. (2014) Agrow – *The Commercial Seed Market in Africa 2014: Countries, Crops and Companies.* Agribusiness Intelligence/Informa, London.

SourceWatch (2012) AGRA's Programme for Africa's Seeds Systems. The Center for Media and Democracy, Madison, Wisconsin. Available at: http://www.sourcewatch.org/index.php/AGRA%27s_Programme_for_Africa%27s_Seeds_Systems (accessed 2 May 2017).

The White House (2012) *Fact Sheet: G-8 Action on Food Security and Nutrition, May 18, 2012.* The White House, Washington, DC. Available at: https://www.whitehouse.gov/the-press-office/2012/05/18/fact-sheet-g-8-action-food-security-and-nutrition (accessed 1 May 2017).

World Bank (2009) *Awakening Africa's Sleeping Giant: Prospects for Commercial Agriculture in the Guinea-Savannah Zone and Beyond.* The World Bank, Washington DC. Available at: http://siteresources.worldbank.org/INTARD/Resources/sleeping_giant.pdf (accessed 1 May 2017).

World Bank (2015) Agriculture & Rural Development Indicators (2015) GDP per capita growth (annual %). Available at: http://data.worldbank.org/indicator/NY.GDP.PCAP.KD.ZG (accessed 1 May 2017).

## Further Resources

African Union (2014) *Science, Technology and Innovation Strategy for Africa 2014.* African Union Commission, Addis Ababa. Available at: http://www.hsrc.ac.za/uploads/pageContent/5481/Science,%20Technology%20and%20Innovation%20Strategy%20for%20Africa%20-%20Document.pdf (accessed 2 May 2017).

African Union (2015) *Agenda 206 – The Africa We Want,* 2nd edn, Popular version. African Union Commission, Addis Ababa. Available at: http://archive.au.int/assets/images/agenda2063.pdf (accessed 2 May 2017).

AGRA (2014) *Africa Agriculture Status Report 2014: Climate Change and Smallholder Agriculture in Sub-Saharan Agriculture.* Alliance for a Green Revolution in Africa (AGRA), Nairobi.

Beintema, N. and Stads, G.-J. (2014) *Taking Stock of National Agricultural R&D Capacity in Africa South of the Sahara. ASTI Synthesis Report.* Agricultural Science and Technology Indicators (ASTI), International Food Policy Research Institute (IFPRI), Washington, DC. Available at: https://www.asti.cgiar.org/pdf/AfricaRegionalReport2014.pdf (accessed 2 May 2017).

Boettiger, S., O'Connor, A., Mabaya, E., Anthony, V., Sperling, L., Barker, I., Le Page, L., Haile, S. and Kuhlmann, K. (2015) *Growing Smartly: Scaling Up Seed Systems for Smallholder Farmers.* Syngenta Foundation for Sustainable Agriculture, Basel, Switzerland, Global Access to Technology for Development (GATD), Berkeley, California with AgPartnerXChange (APXC, http://www.apxchange.org/).

Chambers, J.A., Zambrano, P., Falck-Zepeda, J., Gruère, G., Sengupta, D. and Hokanson, K. (2014) *GM Agricultural Technologies for Africa: A State of Affairs.* Report of a study commissioned by the African Development Bank. International Food Policy Research Institute (IFPRI), Washington, DC. Available at: http://www.ifpri.org/sites/default/files/publications/pbs_afdb_report.pdf (accessed 2 May 2017).

Hickey, L. (2014) The Speed Breeding journey: from garbage bins to Bill Gates. QAAFI Science Seminar, 28 October 2014, Queensland Alliance for Agriculture and Food Information. YouTube video available at: https://www.youtube.com/watch?v=5tsor4PuMmw (accessed 2 May 2017).

ISAAA (2014) *Global Status of Commercialized Biotech/GM Crops: 2014*. ISAAA Brief 49, International Service for the Acquisition of Agri-biotech Applications. Available at: http://www.isaaa.org/resources/publications/briefs/49/executivesummary/default.asp (accessed 2 May 2017).

Kahn, K. and O'Connor P. (2014) *New Product Forecasting and Risk Assessment: How to Deliver Meaningful Numbers for Business Cases, Gate Reviews and Portfolio Management*. Webinar from Sopheon, Bloomington, Minnesota. Available at: https://www.sopheon.com/new-product-forecasting-risk-assessment/ (accessed 2 May 2017).

Keyser, John C. (2013) *Opening Up the Markets for Seed Trade in Africa*. Africa Trade Practice Working Paper Series No. 2, World Bank, Washington, DC. Available at: http://documents.worldbank.org/curated/en/2013/10/18444842/opening-up-markets-seed-trade-africa (accessed 2 May 2017).

Koebner, R. and Fulton, T. (2017) Plant Breeding: Concepts and Methods. A Learning Module. Available at: https://www.integratedbreeding.net/courses/plant-breeding-concepts-and-methods/index-id=002.php.html (accessed 2 May 2017).

KPMG (2013) *The Agricultural and Food Value Chain: Entering a New Era of Cooperation*. KMP International Cooperative. Available at: https://home.kpmg.com/content/dam/kpmg/pdf/2013/06/agricultural-and-food-value-chain-v2.pdf (accessed 2 May 2017).

Kyler, J. (2003) Assessing Your External Environment: STEEP Analysis, MBA Boost – STEEP Analysis Tool Competia.com, Issue 33, December 2002–January 2003. Content available at: https://www.mbaboost.com/steep-analysis-tool/ (accessed 2 May 2017).

Mba, C., Guimaraes, E.P. and Ghosh, K. (2012) Re-orientating crop improvement for the changing climatic conditions of the 21st century. *Agriculture and Food Security* 1: 7. Available at: http://www.agricultureandfoodsecurity.com/content/1/1/7 (accessed 2 May 2017).

Palacios, A.C. (2012) *Drivers of Change: Agricultural Modernization and Women's Status in SAT [the Semi-Arid Tropics of] India*. ICRISAT Working Paper Series No. 35, International Crops Research Institute for the Semi-Arid Tropics, Patancheru, Andhra Pradesh, India. Available at: http://oar.icrisat.org/6502/1/DiversofCHange_WPS_35_2012.pdf (accessed 30 August 2017).

Petrobelli, C. and Puppato, F. (2015) *Technology Foresight and Industrial Strategy in Developing Countries*. UNU-MERIT Working Paper #2015-016, United Nations University–Maastricht Economic and Social Research Institute on Innovation and Technology, Maastricht, The Netherlands. Available at: https://www.merit.unu.edu/publications/wppdf/2015/wp2015-016.pdf (accessed 2 May 2017).

Robinson, S., Mason-d'Croz, D., Islam, S., Sulser, T.B., Robertson, R.D., Zhu, T., Gueneau, A., Pitois, G. and Rosegrant, M.W. (2015) *The International Model for Policy Analysis of Agricultural Commodities and Trade (IMPACT): Model Description for Version 3*. IFPRI Discussion Paper 01483, International Food Policy Research Institute, Washington, DC. Available at: http://ebrary.ifpri.org/utils/getfile/collection/p15738coll2/id/129825/filename/130036.pdf (accessed 2 May 2017).

Schwartz, P. (1996) *The Art of the Long View: Paths to Strategic Insight for Yourself and Your Company*. Random House, New York.

Tadele, Z. (2014) Role of crop research and development in food security of Africa. *International Journal of Plant Biology Research* 2(3): 1019. Available at: https://www.jscimedcentral.com/PlantBiology/plantbiology-2-1019.pdf (accessed 2 May 2017).

## Web resources

AGRA – Alliance for a Green Revolution in Africa, Nairobi. Growing Africa's Agriculture: 10 Years of Success. Available at: http://agra-alliance.org/what-we-do/program-for-africas seed-systems/ (accessed 2 May 2017).
ASTI – Agricultural Science and Technology Indicators. African Country Factsheets. Available at: http://www.asti.cgiar.org/countries (accessed 2 May 2017).
CropLife International – Database of the Safety and Benefits of Biotechnology. Available at: http://biotechbenefits.croplife.org/ (accessed 2 May 2017).
FAO – Country Profiles. Country-specific information from the entire FAO website with a direct link to the country website for further information. Available at: http://www.fao.org/countryprofiles/en/ (accessed 2 May 2017).
FAO – FAOSTAT database on agricultural production and trade by crop, country and year. Available at: http://www.fao.org/faostat/en/#data (accessed 2 May 2017).
FAO – Food Balance Sheets: per capita consumption by food type by country and region. Available at: http://www.fao.org/faostat/en/#data/FBS (accessed 2 May 2017).
International Trade Centre – International trade statistics. Available at: http://www.intracen.org/itc/market-info-tools/trade-statistics/ (accessed 2 May 2017).
ISAAA – International Service for the Acquisition of Agri-biotech Applications. GM Approval Database of Biotech/GM crop approvals featuring Biotech/GM crop events and traits that have been approved for commercialization and planting and/or for import for food and feed use, with a short description of the crop and the trait. Available at: http://www.isaaa.org/gmapprovaldatabase/default.asp (accessed 2 May 2017).
Monsanto – Monsanto's Wheat Platform. Available at: http://www.monsanto.com/products/pages/monsantos-wheat-platform.aspx (accessed 2 May 2017).
ReSAKSS – Regional Strategic Analysis and Knowledge Support System. African CAADP (Comprehensive Africa Agriculture Development Programme) Country Investment Plans. Available at: http://www.resakss.org/publications/aw?key=&type=CAADP+Country+Investment+Plans&country=Africa+wide&topic=0 (accessed 2 May 2017).
SeedQuest Africa – Information and news for African seeds from this global information service for seed professionals Available at: http://www.seedquest.com/news.php?id_category=444&id_region=11 (accessed 2 May 2017).
World Bank Development Indicators (WDI). Available at: http://wdi.worldbank.org/tables (accessed 2 May 2017).
World Bank Population Indicators. Available at: http://data.worldbank.org/indicator/NV.AGR.TOTL.ZS/countries (accessed 2 May 2017).

# 3 Understanding Clients' Needs

Pangirayi Tongoona*, Agyemang Danquah and Eric Y. Danquah

*West Africa Centre for Crop Improvement (WACCI), College of Basic and Applied Sciences, University of Ghana, Legon, Ghana*

## Executive Summary and Key Messages

### Objectives

**1.** To equip plant breeders with the knowledge and methods to understand clients and their value chains, their needs, what they prefer and what they are prepared to pay for in a new variety.
**2.** To understand markets and the need, importance and principles of market research and best practices to guide the information gathering from clients and from crop value chains to drive and validate new variety designs.

The chapter aims to enables plant breeders to: (i) define clients and stakeholders; (ii) understand the various categories of clients in value chains – including seed distributors, farmers, processors, traders, retailers, marketers and consumers, as well as their activities; (iii) identify market segments and their importance in determining the number of new varieties required; and (iv) understand different types and methods of market research and the best practices to follow in order to obtain the information required to design 'fit-for-purpose' new varieties from clients and stakeholders.

### How does demand-led breeding add value to current breeding practices?

- **Client focus.** Breeding goals and objectives are set based on what clients want and need without bias towards either what technology can offer or a specific focus on individual trait improvement.

*Corresponding author: E-mail: ptongoona@wacci.edu.gh

- **Value chains.** Greater understanding is required about the structure of crop value chains, and the buying and selling factors of different clients and their relative priority in setting new variety designs.
- **Participatory approaches.** Demand-led breeding includes, but goes beyond, farmer participatory breeding. It puts more emphasis on regularly consulting and understanding the needs and preferences of all clients and stakeholders in a crop value chain, including, but not restricted to, farmers. It also seeks information from consumers in urban locations and rural areas through participatory rural and urban appraisals. Consultation is a continuous requirement throughout the whole process of variety development, registration and launch, as well as post release.
- **Dual-purpose varieties.** A new variety not only supports farmers' requirements for crop productivity and home consumption but also ensures that surplus crop produce can enter markets, thus generating cash returns to all of the value chain participants.
- **Variety design and benchmarking.** Stronger emphasis is placed on systematic, quantitative assessment of varietal characteristics and on creating product profiles with benchmarks for varietal performance and line progression. Consumer-demanded traits are recognized as being as important as production traits. This requires a greater strategic prioritization of traits among the many that are required for farmers, processors, seed distributors, transporters, retailers and consumers. Hence, it may involve the development of different varieties for different segments of the value chain.
- **Market research.** Stronger emphasis is given to gathering unbiased, reliable, independent information on clients' needs and preferences.
- **Market and business knowledge.** Breeders require greater knowledge about crop uses, markets and the 'business/economics' of breeding. This takes the best practices from plant breeding and integrates them with the best practices in business.

*Implications for the role of the plant breeder*

- **The business of plant breeding.** Breeders are required to learn the 'business of plant breeding' and be able to gather and assimilate information from multiple clients and sources within crop value chains. The approach uses participatory appraisals in both rural and urban situations, including devising participatory appraisal methods to interact with traders, retailers and consumers in urban situations.
- **Partnering and collaboration.** Breeders can broaden their reach, influence and know-how beyond their technical competencies and so be able to manage collaborations and partnerships with the private sector, policy makers and investors.

## Key messages for plant breeders

*Crop uses*

- An understanding of all crop uses and of the effect the properties of a variety have on individual crop uses is vital to successful demand-led breeding and high market adoption rates.

### *Clients and stakeholders*

- A *client* is a customer, buyer, purchaser or receiver of a new crop variety, its crop produce or processed material from a seller, vendor or supplier in the value chain for a monetary or other consideration.
- The term *stakeholder* is broader and includes clients and any other person or organization with an interest in a given crop situation, action or enterprise.

### *Market segments*

- A clear definition is required of clients with similar requirements as these then represent *market segments*. Product profiles need to be created for specific market segments of clients.

### *Clients within value chains*

- **Understanding clients.** An understanding of clients is central to demand-led variety design, release and adoption. It is essential to be clear on who the clients are and what affects their buying decisions.
- **Value chains.** Breeders need to understand value chains and the relative importance of different clients in the chain, as well as their requirements in new variety design.
- **Different clients.** Different clients in value chains have different requirements and not all of these requirements can always be satisfied with the same variety, especially when there are specialist properties required for processing. Breeders should have regular contact with clients in all parts of the value chain and involve them in new variety design.
- **Client location and scale.** The geographic location of clients is important, as is knowledge of whether the benefits and value of new varieties are also applicable for potential clients across national borders. The analysis of agro-ecological zones should be given particular attention. The more clients that can benefit from each new variety, the greater the investment case for a breeding programme, especially when it can have a multi-country impact.

### *Understanding markets and market research*

- **Best market research practices.** Breeders need to access key information from farmers and all clients in crop value chains. The best quality market research methods – those that remove bias – should be used. This requires skilled third parties, such as social scientists and independent market research agencies, to gather information and conduct interviews on behalf of their plant breeder clients.
- **Partnering with social scientists.** Breeders should liaise closely with the social scientists who are conducting *ex ante* impact assessments and seek to include additional questions that can be used to influence new variety design and plant breeding programmes.
- **Participatory appraisals.** Participatory rural appraisals (PRAs) are a starting point for gathering market research information. PRAs focus on farmers, but they also need to be extended to many other clients in the value chain,

especially to traders, retailers and consumers in urban and city situations. PRAs then become 'participatory appraisals'.
- **Avoid making assumptions.** Breeders require regular contact with key clients in order to avoid making incorrect assumptions based on the use of historical data rather than on the current situation and future projections.

### Key messages for research and development (R&D) leaders, government officials and Investors

*Clients within value chains*
- **Understanding clients and value chains.** More support is needed, not only for technology and scientific capacity building in breeding programmes in Africa, but also for assessing the needs of clients and value chains.
- **Seed system development.** For the development of seed systems and for improved seeds to reach farmers, especially in remote locations, distributors require portfolios of 'fit-for-purpose' varieties. Portfolios of new varieties are also required for market creation, growth and business sustainability.
- **Public–private sector partnerships.** In the first instance, public sector breeding programmes should be the key source for clients and value chains of new varieties of food security crops that are currently not commercial (export) crops. In the longer term, a developing local private sector seed business is a more sustainable strategy for both food security crops and export crops.

*Understanding markets and market research*
- **Markets and segments.** Enabling breeders to understand markets and market segments requires endorsement and support from R&D leaders, managers, government officials and investors.
- **Market research.** R&D leaders, managers, government officials and investors need to recognize that detailed information is required from clients throughout a whole crop value chain. This requires adequate funding for market research to be included both in national R&D budgets and in project proposals for external grants. Social scientists are well equipped to undertake market research, and during impact appraisals they should be requested to include additional questions on topics that are relevant to plant breeders.

## Introduction

The objectives of this chapter are:

**1.** To equip plant breeders with the knowledge and methods to understand clients and their value chains, their needs, what they prefer and what they are prepared to pay for in a new variety.
**2.** To understand markets and the need, importance and principles of market research and best practices to guide the information gathering from clients and from crop value chains to drive and validate new variety designs.

The aim of the chapter is to enable breeders to: (i) define clients and stakeholders; (ii) understand the various categories of clients in value chains – including seed distributors, farmers, processors, traders, retailers, marketers and consumers, as well as their activities; (iii) identify market segments and their importance in determining the number of new varieties required; and (iv) understand different types and methods of market research and the best practices to follow in order to obtain the information required to design 'fit-for-purpose' new varieties from clients and stakeholders. It also aims to act as a resource for education in this field. Towards achieving this purpose, boxes are included in several sections of the chapter that summarize the educational objectives of these sections and present the key messages and questions that are involved. There is also a final box at the end of the chapter that summarizes the overall learning objectives.

## Crops and their Uses (Box 3.1)

Plant species have evolved over the millennia, with vascular plants first occurring on the planet around 130 million years ago. *Homo sapiens* is estimated to have appeared less than 500,000 years ago. Since that time, humans have capitalized on nature's diversity and bounty for food, clothing, medicine, energy and shelter, as well as for the air that we breathe. Most of the 330,000 plants on the planet are inedible as a result of their chemical defence arsenal against predators, fungi, bacteria, viruses and nematodes. Only about 7000 plant species or their parts comprise food sources and, of these, about a dozen are major food crops.

We are resourceful with the plants around us and the crops we grow. Typically, we think of the specific part of a plant that we eat, without paying much attention to the rest of the vegetative biomass, but in places where resources are limited, every part of the crop is utilized as food, animal feed, shelter, bedding, etc. So we need to view the whole crop plant and take a holistic approach to understand the social and community norms. In demand-led breeding, it is vital to understand all of the uses of the crop plant, and the people, clients and markets that may be affected by changing specific plant characteristics. This includes considering crops that require processing or cooking to become food. At the simplest level, this can be hand milling sorghum or wheat to make flour, or pounding and boiling cassava to denature residual toxic cyanogenic glucosides. A key component to successful demand-led variety design is to understand all of these processing and cooking components, and to incorporate these requirements into new designs and the process of selecting for the best characteristics. As an example, Fig. 3.1 illustrates the range of fresh and processed food uses of tomato in Ghana.

Crops have many uses that go beyond their use as food, such as their use as animal feed, in industry, for energy and in public health and medicines. For example, cassava (*Manihot esculenta*) is an important staple food crop in the humid and subhumid zones of sub-Saharan Africa, where it provides over half of the dietary calories for over 200 million people. It was introduced into central

**Box 3.1.** Crops and their uses: educational objectives.

**Purpose:** to understand the importance of identifying all uses of a crop and the range of clients and stakeholders that should influence the design of new varieties.

**Educational objectives:**

- to understand the diversity of crop uses;
- to be able to define and profile the clients(s) for a particular breeding programme and their needs, and who and what new varieties are being designed for; and
- to recognize there are different people who influence that breeding programme, and have a clear understanding of each of their roles and influence.

**Key message**

- Understanding all crop uses and the effect that the properties of a variety have on individual crop uses is vital to successful demand-led breeding and high market adoption rates.

**Key Questions**

- Who will use your variety and for what purpose?
- How would you define a client, a customer and a stakeholder?
- How many clients or client categories do you have?
- How many farmers can your variety serve?
- What are their needs?
- Where are they located?
- Who has dominance in the market?
- In relation to farmers, are clients in the same or different agro-ecological zones?
- Are there farmers with similar needs located in the similar ecological zones in neighbouring countries or regions who could also be clients for your varieties?
- What are the demographics of the clients?
- Recognize there are multiple clients with different needs (e.g. farmers, traders, industry and the consumer) and that these needs may be similar or different in various parts of your country and in neighbouring countries. What are these needs and where are they located?

and West Africa from Brazil by Portuguese traders in the 16th century, and then spread widely across Africa. In addition to its widespread use as a starchy staple food in Africa and Latin America, cassava also has multiple other uses, including as a source of starch and alcohol for industrial purposes, and as processed animal feed. In Asia, cassava is primarily grown as an industrial crop rather than a food crop. These multiple purposes need consideration by breeders when they are designing new varieties (see Table 3.1).

The reason why it is important to understand all crop uses, client requirements and social and farming customs is because there are many examples of cases where a lack of knowledge or inattention by breeders to these matters has meant that breeding goals were set that led to 'improved' varieties that were not adopted by farmers (see Chapter 1, this volume, on variety adoption).

An example is the introduction of dwarfing genes into wheat and rice as the genetic basis of the 'Green Revolution', which transformed wheat and rice

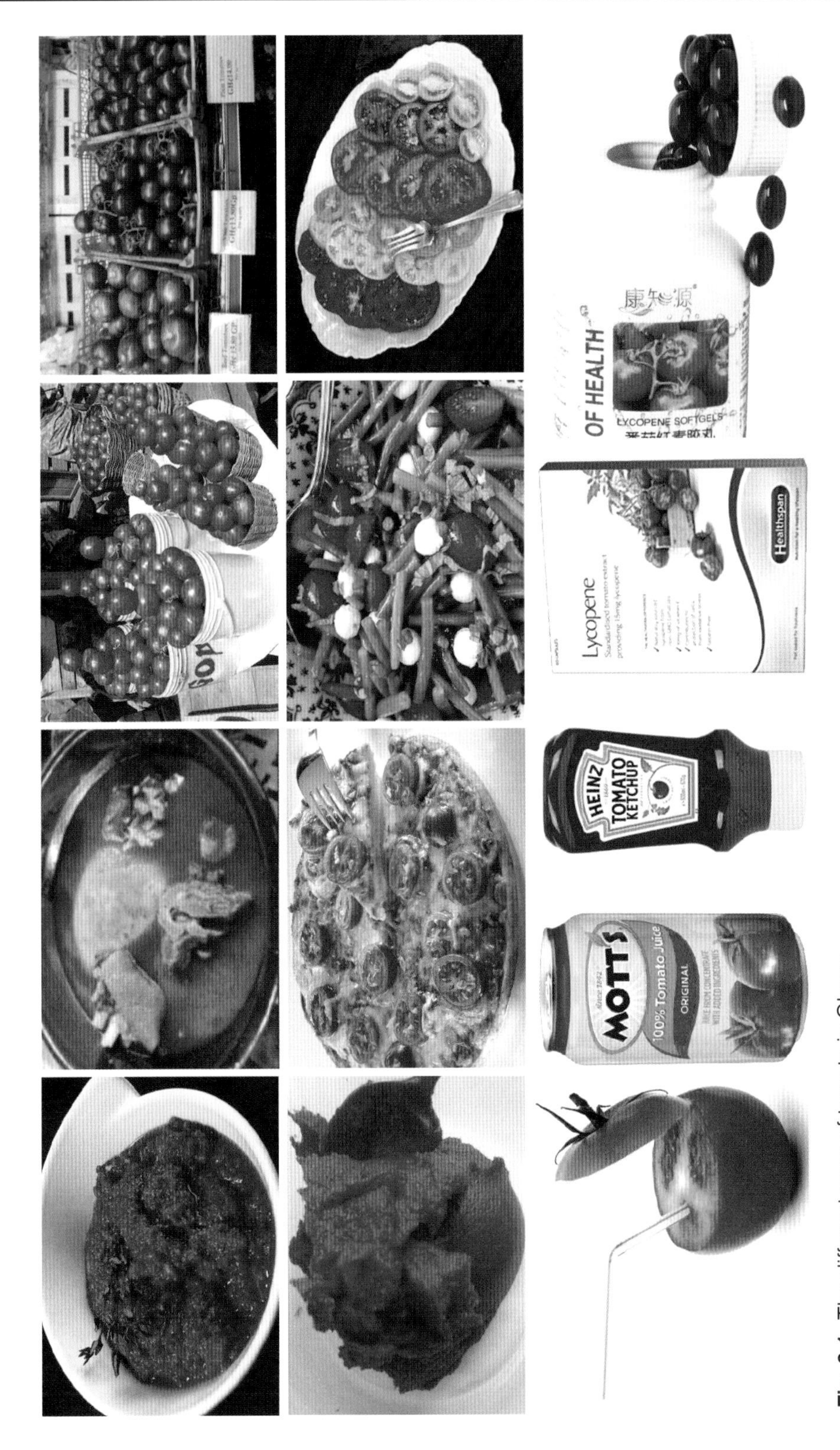


**Fig. 3.1.** The different uses of tomato in Ghana.

**Table 3.1.** Cassava as an example of a multipurpose crop. Data from Dziedzoave and Hillocks (2012) and Kleih *et al.* (2013).

| Type of market | Uses |
|---|---|
| Food | Pounded to flour, boiled or mashed<br>Fermented or made into dried chips |
| **Other markets** | |
| Animal feed | Poultry, pigs, sheep and goats |
| Beer | Starch source for alcohol production |
| Industrial alcohol | Starch source |
| Laundry industry | Starch source |
| Mosquito coils | Starch source |
| Paper | Glue adhesive |
| Plywood | Glue extender |

production in the Americas and Asia, respectively, from the 1970s onwards. A similar genetic approach was tried with cereals in Africa, whereby breeders introduced dwarfing genes into tall sorghum lines in West Africa and Ethiopia on the premise that the new varieties would produce substantial improvements in grain yield similar to those obtained in dwarf varieties of wheat and rice. However, these short straw varieties were not adopted by farmers in Africa because the strong, tall stalks of the original tall sorghum lines are important for use as housing and fuel.

Similarly, rice farmers in Mali preferred tall rice varieties because the harvesting of short rice varieties became difficult for women with young children on their backs when they had to bend down for long periods of time (Effisue *et al.*, 2008).

Breeders in the search for yield and pathogen resistance have tended to place less importance on consumer and processing attributes, and as a result, local varieties often remain preferred to new, 'improved' varieties. An example is the higher yielding, cassava mosaic disease (CMD)-tolerant varieties of cassava bred in Zambia and Malawi. Adoption rates were low and slow during 1990–2008 because the new varieties lacked the consumption attributes that were highly valued by farmers (Alene *et al.*, 2013). Similarly, Timu *et al.* (2014) highlighted the importance of consumer attributes such as taste, brewing quality and ease of cooking for sorghum in their study investigating how multiple traits affected the varietal adoption of sorghum in Kenya.

Another important aspect for breeders to consider is the source of the genetic variation available that reflect of the range of characteristics required and might be used to create potential improvements and new market segments. This is because the greatest genetic diversity tends to be located near to the 'centres of origin' of crop species. African diets include many crops that originated on other continents, including such staple foods as cassava, maize, potatoes and tomatoes, all of which originated in the Americas. Africa also has its own rich biological heritage, and this includes many crops that have their centre of origin in Africa, including finger millet, teff (*Eragrostis tef*), pearl millet, African rice, sorghum, pigeon pea, bambara groundnut (*Vigna subterranea*), watermelon, musk melon, coffee, African eggplant (aubergine) and oil palm. These crops have since spread out of Africa to other continents via long-standing trade routes, similar to the trade routes that also introduced crops into Africa from other continents.

Continuing international exchange of crop germplasm across countries and continents is of mutual benefit for crop improvement programmes worldwide when seeking genetic diversity to address biotic and abiotic stresses, as well as consumer-preferred traits.

## Clients and Stakeholders (Box 3.2)

The terms 'client' and 'stakeholder' are in common use in the business of plant breeding and new variety development. A client is a customer, buyer, purchaser or receiver of a new crop variety, its produce, or processed material from a seller, vendor or supplier in the value chain, for a monetary or other consideration. Within the context of demand-led breeding, clients include seed

**Box 3.2.** Clients and stakeholders within value chains: educational objectives.

**Purpose:** a clear understanding of the types of individuals and organizations that are in value chains; and also of their different activities, requirements for raw materials and relative importance in contributing to new variety design.

**Educational objectives:**

- to understand the concepts of a value chain;
- to understand the clients and stakeholders involved in the value chain;
- to define the value chain(s) for a breeding programme;
- to know the right people in the value chain and create a multi-stakeholder platform for engagement and collaboration or resource mobilization; and
- to be able to assign the relative importance of the different stakeholders to the design of the variety being developed in the breeding programme.

**Key messages**

- **Understanding clients.** An understanding of clients is central to demand-led variety design, release and adoption. It is essential to be clear on who the clients are and what affects their buying decisions.
- **Value chains.** Breeders need to understand value chains and the relative importance of different clients in the chain as well as their requirements for new variety designs.
- **Different clients.** Different clients in value chains have different requirements and, often, all clients cannot be satisfied by the same variety, especially when there are specialist properties required for processing. Breeders should have regular contact with clients in all parts of the value chain and involve them in new variety design.
- **Client location and scale.** The geographic location of clients is important, and also whether the benefits and value of new varieties are applicable to potential clients across national borders. The analysis of agro-ecological zones should be given particular attention. The more clients that can benefit from each new variety, the greater the investment case for a breeding programme, especially when it can have multi-country impact.

*Continued*

**Box 3.2.** Continued.

**Key questions**

- What is a crop value chain?
- What is the crop value chain for your plant breeding programme?
- Who are the clients and stakeholders in the value chain?
- What is their business (e.g. growing, processing, wholesaling or exporting)?
- What is the relative importance of each client in your value chain?
- How do you engage with all the clients in the value chain of your product?
- How do you prioritize and weight client needs (e.g. by size of market, profitability or ease of supply)?

companies, farmers, crop traders, processors, food companies, retailers and consumers. Stakeholder is a broader term that includes clients, but also any person or organization with an interest in a given crop situation, action or enterprise. It includes anyone who can internally or externally affect, or can be affected by, an organization, strategy or plant breeding programme that is creating new varieties. Thus, stakeholders can also include government officials, investors and civil society organizations, among others.

Estimating the number of farmers who will adopt a new variety is a core element of demand-led breeding. Their purchasing power drives investment cases for plant breeding, as farmers are the suppliers of produce to meet demand and the key contributors to achieving food security.

Most farmers in Africa grow crops for food and to improve their livelihoods by selling surpluses into their local, national and regional markets. If they are growing cash crops and are well organized, they may also enter international markets, such as those for coffee, cocoa, cashews and sesame. They are likely to adopt a new variety if it has qualities for ease of cultivation, harvesting and postharvest processing (is easy to cook, has a good taste and other culinary qualities) that are required by their markets. Their decisions are influenced by many social and economic parameters, not least that the overall benefits of changing varieties outweigh the risks.

## Markets and Market Segments

### Key message

- Clients with similar requirements need to be clearly identified as these groups represent market segments. Product profiles need to be created for specific market segments of clients.

'Market segmentation' can be defined as: 'A strategy that involves dividing a broad target market (such as a crop) into subsets of consumers, businesses or countries with common needs and priorities, and then designing and implementing strategies to serve them'.

A new crop variety should be designed to meet the needs of a group of individual clients or organizations that have a set of common requirements. This group is termed a 'market segment'. An understanding of market segmentation is required because it defines the scope and likelihood of farmer adoption of a new variety. It is rare that a single crop variety design will be able to meet all the requirements of farmers or consumers in a specific country or region. This is because there are many differing climatic, environmental, agro-ecological, commercial and logistical requirements for that variety design. Agro-ecological zones require particular attention here because modern satellite, geological and climatic data can help breeders to map out the potential users of their prospective varieties. Agro-ecological zones have no political boundaries and may occur in the same or different, neighbouring countries. Therefore, it is vital to understand these zones and the number of segments of clients with similar requirements, and thus the number of variety designs that are required.

A market segment always has a monetary or social value that forms the basis of an investment case for a breeding programme. Each market segment has a specific size and monetary value that requires analysis of the clients purchasing the variety, e.g. the number of farmers, their location, their expenditure on seed and inputs or, potentially, what they can or would pay for an improved variety.

Examples of market segments in different parts of a value chain are: (i) a group of farmers that grow a crop in a particular agronomic way (e.g. tomatoes grown under open field conditions in a specified agro-ecological zone in a country; or tomatoes grown in intensive production under vinyl); (ii) traders selling a specific market class of beans; (iii) a group of processors that requires a unique type of cassava that is suitable for the production of alcohol for fuel; and (iv) consumers with different varietal preferences, e.g. the purchase of Irish potatoes for either boiling or frying.

### Case study: tomato market segmentation in Ghana

There are three segments of retailers selling tomatoes in Ghana: local markets, roadside groceries and supermarkets. There are also three types of purchasers: hotels and restaurants, local households and expatriates (Fig. 3.2). Further, there are different tomato varieties targeted at these different tomato customer segments, respectively: (i) large tomatoes for slicing and cooking; (ii) small cherry tomatoes for salads; and (iii) premium-priced, mini, amber plum tomatoes for snacking (Fig. 3.3).

## Value Chains and Clients Within Value Chains

Subsistence farming, in which farmers save their own seed, do not use any fertilizer or other inputs, and all of the produce is consumed by their own families, means such farmers are essentially self-sufficient. However, once a surplus is produced and farmers buy and sell inputs and outputs, then they are participating in a 'value chain'. Value chains are often complex, with multiple actors

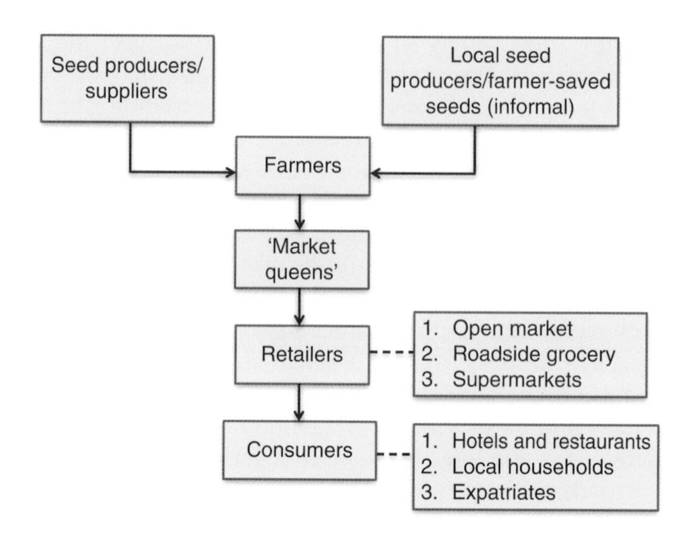


**Fig. 3.2.** Retail and purchaser segments for fresh tomatoes in Ghana ('market queens' oversee entire marketplaces).

| Type: Large tomatoes | Type: Cherry (on the vine) | Type: Mini, amber plum |
|---|---|---|
| Use: Fresh sliced, cooking | Use: Salads | Use: Salads and snacking |
| Pricing: Economy price | Pricing: Mid-range price | Pricing: Premium price |

**Fig. 3.3.** Types of fresh tomatoes, their uses and their pricing segments in Ghana.

buying and selling in parallel, and they may be more aptly described as 'value networks'. Historically, the first value chains started when people built settlements and started to farm. In Africa, this was about 10,000 years ago; breeding also commenced around that time, with farmers saving seed and making their own selections.

A 'value chain' is composed of the activities, services and purchasing decisions that bring a product from its conception to its end use in a particular industry. It comprises groups of clients with similar needs buying and selling either whole crops or processed components from the raw plant materials. The value chain is described by the series of activities and actors along the supply chain, and what and where value is added by these activities and actors. A demand-led breeder needs to understand all of these elements of the value chain of the crop, so that the needs of clients can be included in their new variety designs. Understanding value chains becomes especially important when economic development leads to changing diets, as with the rise of the middle classes, from whom there is greater demand for animal source protein-based foods, fruits and vegetables.

Value chain analysis defines where, how and why value is added and created along the chain. Other than the farmers themselves, a 'value chain actor' is a person or organization buying or selling crop or processed products in the value chain. A 'value chain service provider' is an enterprise, such as a bank, creditor, insurance company or regulator, who provides specialized services to actors in the value chain. In a more complex value chain or network, each actor conducts different activities that add value within the chain. Table 3.2 compares the actors in a simple value chain and those in a more complex value chain network.

In simple crop value chains, seed companies supply seed to farmers and the farmers produce crops that are sold to traders. The traders transport the produce to the processors and local markets which, in turn, supply wholesalers. Wholesalers supply the retailers and supermarkets. From the farmer to the supermarket, value is being added to the primary farm produce at each step. The farmer will grow a new variety that his/her family likes to eat and the traders are able to sell, either at the local market or to the processors. In turn, processors generate products that the wholesalers will buy as they are in demand by retailers and supermarkets because they are sought by consumers (Table 3.2, List 1).

The actors in a more complex value chain are represented in Table 3.2, List 2. An example of such a more complex value chain is that of the bean value chain in Uganda. The activities of all of the clients in this value chain, which supplies beans and bean flour to retailers in Uganda, are illustrated in Figs 3.4 and 3.5.

In other examples, a typical value chain for maize begins with inputs such as seeds and fertilizers at the farming stage and moves on to the production of grain. The farmer can sell the grain to consumers or institutions that will, in turn, mill and process it into flour for consumption or further add value by creating other food products. In a wheat value chain, the variety must have good milling and baking qualities for it to be accepted by farmers, millers, bakers and

**Table 3.2.** Actors in a simple crop value chain and a more complex value chain network.

| Simple crop value chain (List 1) | Complex crop value chain net (List 2) | |
|---|---|---|
| Farmers | Agrochemicals | Millers |
| Local market | Commodity producers | Packers |
| Processors | Consumers | Processed foods |
| Retailers | Crop breeders | Refiners |
| Seed company/distributor | Exporters | Restaurants |
| Supermarkets | Extension services | Seed companies |
| Traders | Farmers | Seed producers |
| Wholesalers | Fast food outlets | Supermarkets |
| | Importers | Variety release agencies |
| | Informal retailers | Wholesalers |
| | Manufacturers | |

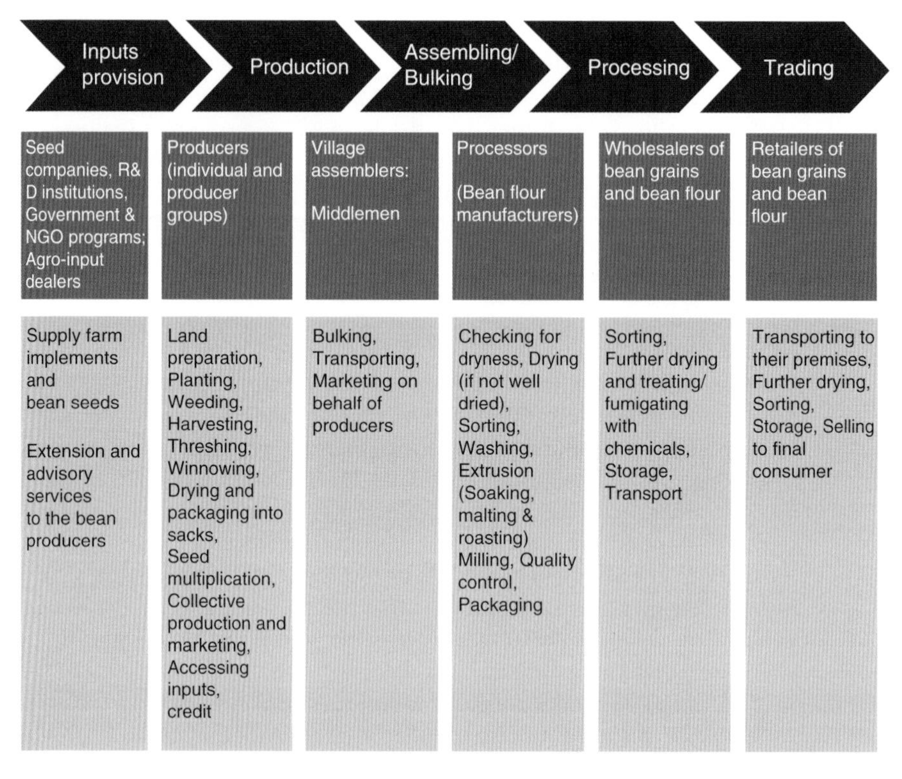


**Fig. 3.4.** Functions of bean value chain actors in Uganda. From the Kilimo Trust (2012), with kind permission.

consumers. The farmer also needs a high yield of the variety because income is maximized according to the quantities sold. Therefore, a high-yielding wheat variety with good milling and baking qualities is ideal for all producers and purchasers in the value chain, and will create demand.

Breeders need to design varieties that will be accepted by all participants in the value chain. This may be possible with a single variety, but market segmentation due to the different requirements of actors in the value chain may mean that different varieties will be required for different purposes. In Ghana, for example, the tomato varieties required for the fresh market are different from those required by the processing market (see Fig. 3.3). The best tomatoes for processing need to have a high Brix value (high sugar content), dense flesh, low water content and low seed number. In contrast, characteristics such as flavour, colour, size and shelf life drive the consumer purchasing of fresh tomatoes.

Typically, farmers and processors should have the greatest role in variety design for staple cereal crops when the baseline germplasm contains the required consumer characteristics for taste and cooking qualities. Yield, productivity and production costs are then the key drivers for design. For fresh fruits and vegetables, the consumer is the prime decider on purchasing, especially in urban situations where shops offer a range of produce to consumers.

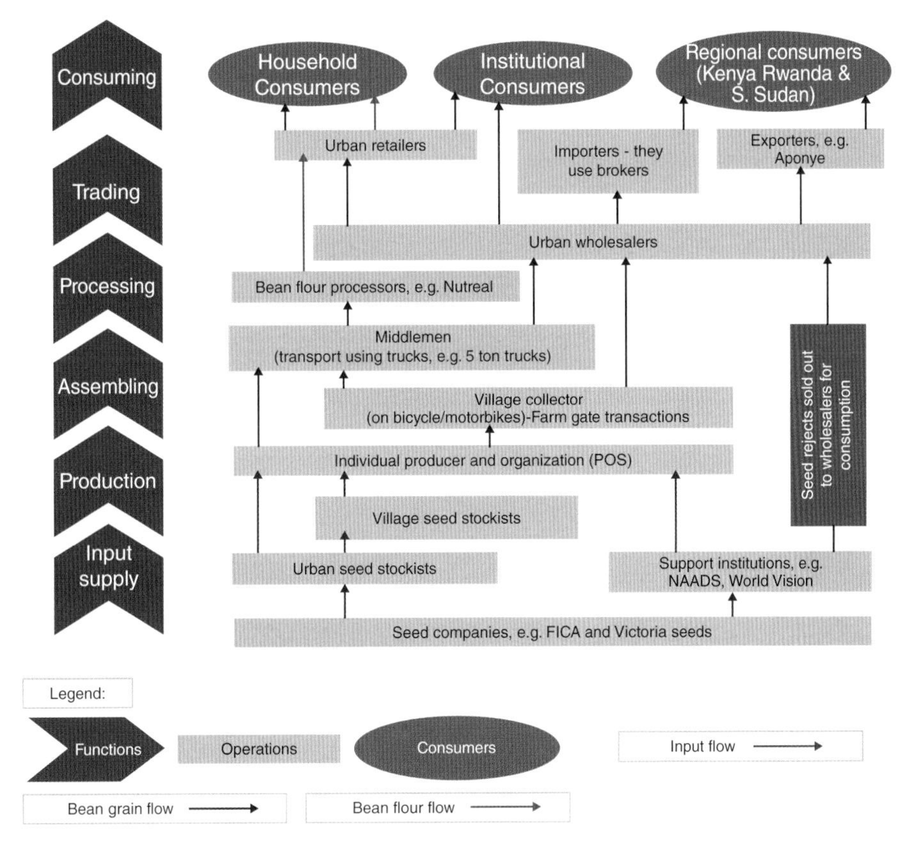


**Fig. 3.5.** Ugandan bean value chain networks. From the Kilimo Trust (2012), with kind permission.

Therefore, consumers should have a greater involvement in determining the product design at an early stage of varietal design for new fruit and vegetable varieties.

## Understanding Markets and Market Research (Box 3.3)

Understanding the market for a new variety is a critical component of successful design. A 'market' is a forum that allows buyers and sellers of a specific good or service to interact in order to facilitate an exchange. The price paid by individuals during the transaction may be determined by many factors, but usually it is driven by supply and demand. Demand-led breeders need to recognize there are different client segments within a market and then commission the appropriate/relevant market research so as to understand the requirements for each.

'Market research' is also termed 'opinion research', 'market survey' or 'poll marketing research'. The first known use of market research was a poll in Delaware in the USA in 1824 (Lockley, 1950). Market research is used in many industries and by public and private organizations, including governments,

**Box 3.3.** Understanding markets and market research: educational objectives.

**Purpose:** to understand markets, and the need, importance and principles of market research and best practices to guide information gathering from clients and crop value chains to drive and validate new variety designs.

**Educational objectives:**

- to understand market opportunities and how they affect new variety design;
- to identify best practices in market research, how to reduce bias and how to decide whether information on customer preferences is reliable;
- to understand that communication is key; and
- to establish a common language among breeders, market research investigators and non-technical colleagues and external advisors.

**Key questions**

- What is a market?
- What types of market exist for your new varieties?
- What kinds of market information do you need?
- What is the size of the total market that your varieties will serve in terms of numbers of farmers and value?
- How many market segments are there and what characteristics are required for each segment?
- What are the information gaps about your markets?
- What is market research and how is it conducted?
- How can you get reliable market information and from whom?
- What are the core principles of best practice in market research?
- Who should be commissioned to do market research?
- When in your plant breeding programme should you consider doing market research to ensure that you maximize the chance of high adoption by farmers and other clients in the value chain?
- What is the difference between market rural appraisals and farmer/consumer market research?
- How can you explore new market information, from whom and from where?
- In addition to market information, who else should be consulted on plant variety design?
- How do you follow changing market demands within your breeding programme?
- How can you respond to the changing markets?

companies, civil society organizations and consumer associations. The most familiar types of market research include traveller surveys at airports, telephone consumer product surveys, online customer satisfaction surveys from product providers and street interviews by pollsters around election time. Market research agencies specialize in best practices and advanced techniques which are aimed at obtaining unbiased, quality data.

International agribusinesses use market research agencies to gather market information from customers about new product requirements and varietal performance. This is so that they gain customer insights that are least affected by existing customer relationships or perceptions. Such agencies provide data of the greatest accuracy and can be trusted to be used for commercial decision making and investment in R&D and plant breeding. Market research is conducted by most private seed companies during new variety development, not just at the product concept stage but also at intervals all the way through breeding, product prototyping, launch and after the commercialization of new varieties.

Public R&D groups and not-for-profit based institutions with a public goods approach tend to focus more on understanding the livelihoods of and decision making by farmers and communities using socio-economic studies by sociologists, maybe in combination with agricultural economists. These studies are often driven by governments or investors wishing to review the impact and benefits of new technology introductions, research investments and/or development interventions. It is unusual for public sector breeders to commission independent market research for the purpose of new variety design. Information is mainly gathered by plant breeders themselves, and this often only at the start of a new plant breeding project.

*Benefits of market research*

Market research enables clients' opinions, needs and wants to be investigated. It can help to identify the needs of all clients in value chains so that solutions are developed that best address these needs. The benefits of market research include:

- ensuring that smallholder farmers can access varieties that will improve their livelihoods;
- enabling farmers not only to produce crops with a higher yield but also crops with the qualities needed to sell into markets;
- setting R&D priorities and delivering varieties that can stimulate market growth and value creation; and
- validating new variety designs, concepts or solutions.

*Risks of not using market research*

There are risks if market research is not used to understand and drive the targets for crop improvement. These risks include the following:

- New varieties, products or technological innovations are created but they are not used by farmers.
- Public or private money is wasted by developing redundant varieties.
- Investment in plant breeding does not give a return on investment and innovation is not sustainable.
- A lack of confidence is generated within government agencies in the performance of their national research systems; this leads to finance ministries seeing R&D as a cost and not as an investment with a return.
- A lack of involvement and lack of credibility and trust by the public arises towards agricultural scientists and the new varieties that they are releasing.

*Quality market research: core drivers and best practices*

For reliable and accurate information there is need to be aware of best practices in conducting market research for demand-led variety development. The best quality market research is based on the following core principles:

- 'Fit-for-purpose' methods are used that gather qualitative and quantitative data, such as:
  - **personal interviews** that involve unstructured-open ended questions;
  - **focus groups** in which a moderator uses a scripted series of questions or topics to lead a discussion among a group of people;
  - **questionnaire** surveys administered to a sample that represents the target market, with the larger the sample the better, as finances permit;
  - **observations** in which videotaping is carried out and observations made of customers' shopping patterns; and
  - **germplasm/prototype testing** that uses trials that involve placing a new product in selected places and testing responses and reactions of customers under real life conditions. Such trials can help to make product modifications, adjust prices, or improve packaging.
- Appropriate communication tools are used, e.g. face-to-face, telephone, Internet, e-mail.
- The methods used are designed to understand the market and clients' needs and the motives driving purchasing decisions, which are termed 'key buying factors'.
- These methods also understand the psychology of respondents and the best ways to phrase questions.
- The risk of bias is minimized by using interviewers who are not from the organization seeking the information, thereby removing errors created as a result of existing human relationships or perceptions.
- The methods are cost-effective.
- They are of adequate scale.
- Time of day of the interview matters and should be carefully considered.
- The length of the interview or questionnaire is important.
- Market research is conducted at appropriate points during new variety development, starting at the concept stage and continuing all the way through the breeding programme, varietal release and after commercialization.

#### *Participatory rural appraisals*

PRAs form a key method for a breeder to gain knowledge and opinions about the lives and needs of rural farmers. The breeder wants to discover traits that farmers prefer in a variety so that these are included in the design of that variety. Varieties developed that incorporate the identified traits should increase adoption by farmers. PRAs typically involve personal interviews, focus groups and questionnaires. They are a core source of market research information. PRAs are also a well-recognized and respected approach used by civil society organizations and other implementing agencies in the planning and management of development projects and programmes. Breeders need to extend the concepts of PRAs to include the needs and views of urban consumers, as part of their formal information gathering process.

Socio-economic studies are also a useful source of information, especially before breeding programmes commence. However, *ex ante* impact studies often do not contain questions that are specifically worded to support plant breeders with their programmes; these questions are designed more to assess the impact of technology or other development interventions. Demand-led breeders should, therefore, be aware of the socio-economic studies being organized that involve farmers, potential users and clients, and then seek to collaborate with the principal investigators so that additional questions are included in *ex ante* studies that identify the key characteristics that are important to farmers in their adoption of new varieties.

## Learning Methods (Box 3.4)

Before this chapter concludes, a summary is provided in Box 3.4 of learning methods – together with assignments and assessment methods – for the main topics that have been covered in the chapter: Clients, Market Segments and Crop Uses; Value Chains; and Market Research.

**Box 3.4.** Learning methods, assignments and assessment methods.

**Clients, Market Segments and Crop Uses**

*Learning method*

- PowerPoint presentations on core principles and definitions of *clients*, *stakeholders*, *market segmentation* and the importance of understanding *crop uses*.
- Group discussions on the range of uses of crop plants and the terminology for understanding clients and the composition of different market segments.

*Assignment*

- Participants to define the types, location and needs of the clients and organizations they are targeting in their crop breeding programmes.
- Participants to identify all market segments and the number of different varieties that are needed to serve all clients adequately.

*Assessment*

- Assignment.
- Exam questions on clients, market segments and crop uses.

**Value Chains**

*Learning method*

- PowerPoint slides to understand value chains and Web searching for information on differences in the architecture of value chains.

*Assignment*

- Detailed evaluation of the crop value chain architecture and the composition of the clients and organizations that a participant's breeding programme serves. Also, identification of stakeholders that influence the structure and dimensions of the value chain and explaining why they do.

*Continued*

**Box 3.4.** Continued.

*Assessment*

- Assignment.
- Exam questions on the structure and composition of value chains.

**Market Research**

*Learning method*

- PowerPoint presentations covering the core principles of best practice in client and value chain market research.
- Group discussion on points of best practice when conducting rural participatory appraisals and how to adapt these methods so that they can be used with retailers, consumers and other value chain actors living in urban areas.
- Group discussion on the goals and contents of socio-economic impact studies and how additional information and questions can be included in the studies to serve the need of plant breeders to understand demand.

*Assignment*

- Designing a market and client research investigation for a participant's demand-led breeding programme that will either validate existing assumptions on variety design or will contribute to creating a new variety product profile/variety design.

*Assessment*

- Assignment.
- Exam questions on best practices in market research to create new variety product profiles.

## Conclusion

### How is demand-led breeding different from current breeding practices?

- **Client focus.** Breeding goals and objectives are set based on what clients want and need rather than having any bias towards either what technology can offer or a specific focus on individual trait improvement.
- **Value chains.** Greater understanding is required about the structure of crop value chains and the buying and selling factors of different clients, together with their relative priority in setting new variety designs.
- **Demand-led breeding.** This approach puts more emphasis on regularly consulting and understanding the needs and preferences of *all* clients and stakeholders in a crop value chain, not only farmers.
- **Variety design and benchmarking.** Stronger emphasis is placed on the systematic, quantitative assessment of varietal characteristics and the creation of product profiles, with benchmarks for varietal performance and line progression.
- **Market research.** Stronger emphasis is placed on gathering unbiased, reliable, independent information on clients' needs and preferences.

- **Market and business knowledge.** A demand-led approach requires breeders to have greater knowledge about crop uses, markets and business. It takes the best practices from breeding and integrates these with the best practices in business.

### Implications for the role of the breeder

- **The business of plant breeding.** A demand-led approach require breeders to understand the 'business of plant breeding' and be able to gather and assimilate information from multiple clients and many sources within crop value chains.
- **Partnering and collaboration.** Breeders need to broaden their reach, influence and know-how beyond their technical competencies, and manage collaborations and partnerships with both the public and private sectors.

## Resource Materials

Slide sets are available for this chapter as part of Appendix 3 of the open-resource e-learning material for the volume. These summarize the chapter contents and provide further information. The e-learning material is available at http://www.cabi.org/openresources/93814 and also on a USB stick that is included with this volume.

## References

Alene, A.D., Khataza, R., Chibwana, C., Ntawuruhunga, P. and Moyo, C. (2013) Economic impacts of cassava research and extension in Malawi and Zambia. *Journal of Development and Agricultural Economics* 5, 457–469.

Dziedzoave, N.T. and Hillocks, R. (2012) Cassava: Ghana Status Report 2011 (PY3 Ending). Cited in: Kleih, U., Phillips, D., Wordey, M.T. and Komlaga, G. (2013) *C:AVA – Cassava Adding Value for Africa. Cassava Market and Value Chain Analysis: Ghana Case Study, Final Report, February 2013.* Natural Resources Institute, Greenwich, UK and Food Research Institute, Accra, Ghana. Available at: https://agriknowledge.org/downloads/cn69m4217 (accessed 3 May 2017).

Effisue, A., Tongoona, P., Derera, J., Langyintuo, A., Laing, M. and Ubi, B. (2008) Farmer perceptions on rice varieties in Sitasso region of Mali and their implications for rice breeding. *Journal of Agronomy and Crop Science* 194, 393–400.

Kilimo Trust (2012) *Development of Inclusive Markets in Agriculture and Trade (DIMAT): A Value Chain Analysis (VCA) of the Bean Sub-sector in Uganda.* United Nations Development Programme, Kampala, Uganda. Available at: http://www.kilimotrust.org/Beans%20VCA%20Report%202012.pdf (accessed 3 May 2017).

Kleih, U., Phillips, D., Wordey, M.T. and Komlaga, G. (2013) *C:AVA – Cassava Adding Value for Africa. Cassava Market and Value Chain Analysis: Ghana Case Study, Final Report, February 2013.* Natural Resources Institute, Greenwich, UK and Food Research Institute, Accra, Ghana. Available at: https://agriknowledge.org/downloads/cn69m4217 (accessed 3 May 2017).

Lockley, L.C. (1950) Notes on the history of marketing research. *Journal of Marketing* 14, 733–736. Available at: http://www.jstor.org/discover/10.2307/1246952?uid=3737760&uid=2&uid=4&sid=21104705243867 (accessed 30 August 2017).

Timu, A.G., Mulwa, R.M., Okella, J. and Kamau, M. (2014) The role of varietal attributes on adoption of improved seed varieties: the case of sorghum in Kenya. *Agriculture and Food Security* 3:9.

## Further Resources

### Web resources

#### *Definitions*

Investopia – Markets. Available at: http://www.investopedia.com/terms/m/market.asp (accessed 3 May 2017).

Wikipedia – Customers and clients. Available at: https://en.wikipedia.org/wiki/Customer (accessed 3 May 2017).

Wikipedia – Market research. Available at: https://en.wikipedia.org/wiki/Market_research (accessed 3 May 2017).

Wikipedia – Market segmentation. Available at: http://en.wikipedia.org/wiki/Market_segmentation (accessed 3 May 2017).

Wikipedia – Participatory rural appraisal. Available at: https://en.wikipedia.org/wiki/Participatory_rural_appraisal (accessed 3 May 2017).

Wikipedia – Stakeholders. Available at: https://en.wikipedia.org/wiki/Stakeholder (accessed 3 May 2017).

Wikipedia – Value chains. Available at: https://en.wikipedia.org/wiki/Value_chain (accessed 3 May 2017).

# 4 New Variety Design and Product Profiling

Shimelis Hussein*

*African Centre for Crop Improvement (ACCI), University of KwaZulu-Natal, Scottsville, Pietermaritzburg, South Africa*

## Executive Summary and Key Messages

### Objectives

**1.** To understand the core methods of product profiling used to characterize existing varieties used by farmers and identify the properties important to clients and stakeholders along the value chain in the future.
**2.** To understand how to create new variety designs and set benchmarks to meet client needs.
**3.** To understand how to prioritize a range of traits using demand-led approaches and make decisions on trait trade-offs.
**4.** To understand how to translate a new variety design into a practical breeding programme with clear goals and objectives.

The chapter aims to enable plant breeders to design new crop varieties that will achieve high adoption rates because their varietal characteristics serve the needs and preferences of farmers, processors, consumers and other stakeholders in the crop value chain.

### How does demand-led variety design add value to current breeding practices?

- **Competitor product profiling.** This requires analysis of the characteristics of existing commercial varieties and landraces as grown by farmers, and

*E-mail: Shimelish@ukzn.ac.za

also of their differentiating characteristics at every stage in the value chain from seed production to farmers, processors, transporters, retailers, food companies and consumers.

- **New variety design.** A detailed product profile is created that contains many traits and characteristics (typically more than 40) with performance benchmarks that are used to create breeding objectives. Current practices often focus on a much smaller number of farmer requirements that are well understood but are not discussed or agreed with other stakeholders in the value chain. Demand-led approaches put more emphasis on combining consumption- and consumer-based traits with farmer requirements to drive adoption. A written demand-led product profile provides a core communication tool that summarizes what clients want and provides the strategic goal for the breeding programme.
- **Quantitative benchmarks.** For each trait, a target quantitative benchmark is set for line progression for variety release, rather than following the common procedure of deciding on a defined number of years for annual selection and progressing the best performing lines for varietal registration at the end of this term.
- **Trade-off decisions.** A decision-making process is used that takes into account client needs, technical feasibility and a range of other practical and fiscal considerations. Active and inclusive decision making is core to demand-led breeding. A prioritized list of traits and the final new variety design that is used to set the breeding goals is discussed and agreed with clients and stakeholders before breeding work commences.

*Implications for role of the plant breeder*

- **Variety identity.** A greater depth of understanding is needed of the full range of characteristics that comprise each variety and landrace used by clients. Demand-led breeders need to recognize the key differentiating features and develop quantitative assessment measures to be able to design successful new varieties.
- **Registration process.** At the variety design stage, breeders should reach out to registration officials and understand their processes and requirements. This is to ensure that appropriate evaluation of consumer-based traits is undertaken that will satisfy regulatory requirements. The dialogue needs to be initiated at an early stage, well before varieties are ready for entering the registration process.
- **Consultation and co-ordination.** Greater consultation, coordination time and liaison skills are needed to understand the needs of clients all along the value chain.
- **Communication skills.** Demand-led breeders must be able to present new variety designs to a wide range of clients, non-technical professionals, government officials and investors.

## Key messages for plant breeders

*Variety design*

- **Product profile.** A specific product profile is required for each segment of clients that a new variety is intended to serve. Each product profile comprises a defined set of prioritized traits.
- **Communication.** A consistent format should be used for product profiles so that they are easy to compare and communicate to clients, plant breeders, scientists, managers and other stakeholders.
- **Validation.** Each new product profile should be tested with clients and assumptions made about acceptability validated before major investment is made in a breeding programme.
- **Market research data.** Qualitative and quantitative data from early discussions with farmers and clients in the crop value chain should be used to create product profiles and make decisions on breeding objectives.
- **Adoption tracking.** Breeders should consider at the variety design stage how adoption tracking will be done, e.g. by phenotypic or genotypic markers.
- **Breeding goals.** Validated product profiles that comprise a predefined, integrated and prioritized set of traits should drive the setting of breeding goals and objectives rather than single traits.
- **Forecasting requirements.** Breeders need to decide how long it will take to develop their new variety and then use scenario-based techniques to review the applicability of their designs on this time frame. Where necessary, designs should be adjusted to take account of the time frame (see also Chapter 2, this volume, Visioning and Foresight for Setting Breeding Goals).

*Setting standards*

- **Breeding objectives.** Clear, quantified breeding objectives with performance indicators are essential.
- **Benchmarks.** Each trait in a product profile should be quantified and measurable versus a defined performance benchmark that needs to be achieved to ensure registration and future adoption by farmers. This benchmark is usually based on the performance of a popular variety or landrace.
- **Bioassays.** Performance must be measurable, with 'fit- for-purpose' assays.
- **Variety registration requirements.** The process of variety registration must be understood at the design phase and early discussions held with key officials, particularly when the design includes consumer-based traits, markers for identification/adoption monitoring and other traits requiring performance assessment, e.g. nutrition, seed certification, etc.
- **Advocacy.** Breeders will need to undertake an early advocacy programme with government officials if changes are required in the registration process to take account of future breeding goals and market-led new variety design features that differ from current traits.

- **Seed production and scaling.** Key design parameters are 'How easily can seed multiplication be scaled' and 'What are the associated costs?' These need to be taken into consideration at the variety design stage so that future demand for seed can be satisfied. Seed production costs can make the difference between a variety being commercially viable or not.

### Key messages for research and development (R&D) leaders, government officials and investors

*Variety design*

- **Variety design.** Clients buy seed of new varieties based on a whole spectrum of attributes and quality characteristics as well as yield. Government officials and investors need to recognize this complexity and provide funding for demand-led product profiles that go beyond single issue varieties.
- **In-house versus external laboratory testing.** Managers and investors should encourage their breeders to consider all options for bioassay and trait performance testing, i.e. whether to conduct these assays in-house, or to contract with outside laboratories and/or partner with the private sector for specialized assays if outsourcing is more efficient and cost-effective.

*Setting standards*

- **Product profile review.** R&D leaders and managers should critically review designs and/or arrange peer review to ensure that new designs and breeding objectives are 'fit-for-purpose'. Investors should expect to see *ex ante* evidence of demand for new varieties.
- **Product profile quantification.** R&D leaders, managers, government officials and investors should expect breeders to provide a greater degree of granularity, quantification and communication of the required performance of traits.
- **Variety registration.** R&D leaders and managers should encourage regular discussions and cooperation between registration officials and their plant breeders.

## Introduction

The objectives of this chapter are:

**1.** To understand the core methods of product profiling used to characterize existing varieties used by farmers and identify the properties important to clients and stakeholders along the value chain in the future.
**2.** To understand how to create new variety designs and set benchmarks to meet client needs.
**3.** To understand how to prioritize a range of traits using demand-led approaches and make decisions on trait trade-offs.

**4.** To understand how to translate a new variety design into a practical breeding programme with clear goals and objectives.

The aim of the chapter is to enable plant breeders to design new crop varieties that will achieve high adoption rates because their varietal characteristics serve the needs and preferences of farmers, processors, consumers and other stakeholders in the crop value chain. It also aims to act as a resource for education in this field. For this purpose, boxes are included in several sections of the chapter that summarize their educational objectives and present the key messages and questions that are involved. There is also a final box at the end of the chapter that summarizes the overall learning objectives.

## New Variety Design and Product Profiling (Box 4.1)

### New variety design: key principles

A new variety design should only be created once the target client(s) and the value chain it is being designed for have been clearly defined (see Chapter 3, this volume). Consultation needs to involve all clients and stakeholders in the value chain who are making or influencing key buying and selling decisions of the crop or its products. The full range of uses of the crop in the value chain need to be understood, e.g. crop use as a food for rural household consumption, as a raw material for processing and/or as a seed or propagating material for production and scaling. The core principles of demand-led new variety design include using best practices in farmer-participatory breeding and participatory rural appraisals (PRAs) to gather key information from stakeholders in rural areas; and similar participatory approaches with consumers, processors, retailers and all stakeholders along the whole value chain. Other stakeholders, such as government officials who influence the enabling market environment, also need to be consulted and their views and activities taken into account.

*Targeting clients and identifying market segments*

Defining a clear group of clients that the new variety is targeted to serve (i.e. its market segment) is the first step in new variety design. A new crop variety should always be designed to meet the needs of a targeted group of individuals or organizations that have a common set of requirements. The concept of market segmentation is important in demand-led breeding because it defines the scope and likelihood of farmer adoption of that variety. Owing to many environmental, agro-ecological, commercial and logistical reasons, it is extremely rare that a single crop variety design will be able to meet all of the requirements of farmers and consumers in a specific country or region. Therefore, it is vital to understand the number of segments of clients with similar requirements that exist and hence the number of variety designs that are required. A market segment always has a monetary or social value that forms the basis of an investment case for a breeding programme. Market segmentation can be

**Box 4.1.** New variety design and product profiling: educational objectives.

**Purpose:** to strengthen the approach breeders use to create new variety designs that deliver the needs and preferences of their targeted clients, including understanding how to construct a product profile with performance benchmarks.

**Educational objectives:**

- to understand all the uses of the target crop and the characteristics required for each use;
- to be able to identify market segments of clients and their uses of the target crop;
- to be able to generate a product profile that defines and communicates the ideal characteristics of an improved variety to these market segments of clients and other stakeholders; and
- to understand registration requirements as they relate to market-required features of new varieties.

**Key messages**

- A product profile is required for each prospective new variety that can be communicated to value chain clients, plant breeders, scientists, managers and other stakeholders.
- Each required characteristic must be quantified and measurable versus a benchmark.
- The product profile needs to be tested with customers before major breeding investments are decided and any assumptions made must be validated before significant investment is made in new breeding programmes.
- Up-to-date qualitative and quantitative market research data on clients' needs are essential to making decisions. Data are required from both participatory rural appraisals (PRAs) and from similar participatory urban studies investigating consumer and retailer needs.
- Clients must be central to decision making.
- Early engagement with clients is essential before significant breeding investment has been made and advanced lines are progressing.
- A key aspect in the design of a new variety (often overlooked) is how seed will be multiplied and what the associated cost may be. This cost can make the difference between a new variety being commercially viable and it not being taken up by the seed system due to costs.
- An understanding is required of how to create an advocacy programme with government regulators and, if necessary, of how to adapt current registration requirements to include the assessment of new, market-led design features that offer additional benefits to farmers and consumers.

**Key questions**

- What types of food products and uses does your crop have?
- Who are you designing your new variety for?
- What is a product profile or ideotype?
- What are the detailed characteristics of the leading crop varieties in the markets that your crop breeding programme is serving?
- How do you assess the performance of varieties that are being used by farmers?
- How do you translate market information into a product profile?
- What characteristics should your new varieties have for all clients in the value chain?
- Have you fully considered all aspects of seed production criteria in your design?
- Does the product profile meet national variety registration requirements?
- Are the registration requirements consistent with the market demand?
- If not, can the registration requirements be refined?

defined as: 'A strategy that involves dividing a broad target market such as a crop into subsets of consumers, businesses or countries with common needs and priorities, and then designing and implementing strategies to serve them' (Wikipedia, 2015; see also Chapter 3, this volume).

Each market segment has a specific size and monetary value that requires a definition of the clients who will purchase the variety, e.g. the number of farmers, their location, their expenditure on seed and inputs, and what they are potentially willing to pay for an improved variety. This also requires an understanding of the value of the processed crop, consumer sales and measures such as social and environmental benefits. If no market currently exists, then estimates are needed of the potential value of an improved variety by making and testing a range of assumptions on the economic, social and environmental benefits of that new variety.

*Value chain needs*

Understanding the needs of clients and stakeholders is required along the whole targeted value chain in rural and urban areas for different crop uses. Historically, plant breeding in Africa has focused primarily on the needs of rural farmers and local markets, but the landscape is changing. The highest population growth rates across Africa are now in the cities and other urban areas. Supermarkets and retailing are changing to meet demand from the rising middle classes. Demand-led breeding recognizes these changing dynamics, acknowledging the need to provide new varieties that will contribute to both food security and value addition, as well as having the new characteristics that people in urban environments want to buy (see Chapter 2, this volume).

*Variety identity and descriptors*

Designing new demand-led varieties requires breeders to understand every facet of the crop plant in their care and the differences between varieties. This includes being able to recognize nuances in plant anatomy, physiology and growth behaviour, and the effects of biotic and abiotic factors on plant performance and on seed production. Equally important is knowledge about the crop's postharvest characteristics, including: (i) its suitability as a raw material for processing; (ii) crop behaviour during harvest, storage and transportation; and (iii) all the unique eating qualities of the crop that are preferred by consumers when it is eaten fresh or when it is cooked.

Demand-led breeders must be extremely familiar with the biological descriptors of their crop and be able to identify varieties that are preferred or rejected by farmers, seed producers, processors, consumers and other stakeholders in the value chain. Specifically, breeders need to understand the reasons why certain varieties are preferred (or not preferred), through discussions with each client and stakeholder; they then need to implement a set of biological tests and field trial designs that can differentiate varietal performance. Benchmarking existing varieties and defining their properties is a vital step in the path to setting the performance standards needed to invest in a new breeding programme and/or proceed with registering new varieties.

*Variety awareness and demand*
Regular contact is needed with clients and stakeholders during new variety development, as is involving them in decision making and the testing of new designs so as to ensure awareness, product pull and widespread use of the new variety when it is registered.

## Product profiling

A 'product profile' is the name given to the technical specification of a new variety. This design specification contains a detailed set of technical attributes or traits with quantitative measures. The best product profiles always set a target benchmark for the required performance of each trait. Typically, these benchmarks are set as comparisons versus the performance of existing varieties, or they are expressed on a numeric or photographic scale. Other names used for a new variety product profile include 'product ideotype', 'variety design' and 'product (or variety) specification' (Anthony, 2013).

*Trait prioritization*
Determining the relative priority of different traits in a new variety design is an important process in demand-led variety design. There may be over 40 traits in a new variety design/product profile. Trait prioritization is required to set the goals and objectives of the breeding programme. Decision tree analysis is recommended as a way to conduct a rigorous assessment of trait prioritization and to deal with a range of often conflicting or mutually exclusive factors. The factors that affect the priority given to various traits are discussed in more detail below.

## Client and market importance

The core goal for a demand-led breeder is to create a new variety that meets client demand by either: (i) improving design features within existing varieties; or (ii) providing new benefits that will increase new varietal adoption. The main inputs required for new variety design are:

- market research;
- variety performance;
- details of the traits required; and
- trait prioritization

*Market research*
If breeders are working closely with clients in the crop value chain, then asking questions to understand and list all of their requirements is straightforward. This is often the case when breeders are using participatory breeding methodologies and having regular contact with farmers. However, systematic market research gathering from all players in value chains, as practised by private sector

seed companies, is generally rare in tropical crop breeding programmes. Such rigorous value chain analysis is likely to highlight many variety characteristics that are required by targeted clients or market segments, and it is common within the private sector to have product profiles with over 30 design features.

*Variety performance*

Detailed technical knowledge is required of the attributes and performance of registered varieties and of those that are market leaders or are gaining market share.

*Traits required*

CLIENT AND MARKET IMPORTANCE OF INDIVIDUAL TRAITS

- Does each trait provide a performance improvement or a unique new benefit?
- How different is the trait to that currently available to farmers or value chain clients?
- What is the relative importance of each trait in the product profile?
- Which of the traits drive the inherent value of the new variety?
- Which traits will drive farmer adoption and market share?
- Which traits will farmers pay more for?
- Which traits can grow the total market value?

A primary consideration in demand-led variety design is how to define the inherent and relative value of each trait in the product profile. A trait may provide a performance improvement over existing varieties and thereby gain market share. Traits that deliver improved yield or resistance against critical pests and diseases are often such market share drivers. If the trait is highly differentiated and delivers a novel benefit that is not currently available, it may be possible not only to gain market share but also to gain a price premium. This demand stimulates growth in the total market value and will be of most interest to the private sector. These types of traits are called 'value creators'; they have the potential to change the market when introduced and are often consumer-based requirements.

An example of value creation is the extended shelf life in tomato that was introduced through tomato breeding in the early 1990s and changed the market in Europe. As a result, Spain became a large exporter of tomatoes to north-west Europe (Bai and Lindhout, 2007).

TECHNICAL AND SCIENTIFIC FEASIBILITY OF TRAIT COMBINATIONS. Trait combinations may be difficult or impossible to achieve due to fundamental physiological or genetic reasons, e.g. deploying five major fungal resistance genes may depress yield.

LEGAL CONSIDERATIONS. Access to germplasm and traits that have intellectual property rights and are owned by others requires careful consideration. Access may not be permitted or royalty payments may be required. Also, international legislation, such as phytosanitary laws, the International Treaty on Plant Genetic Resources for Food and Agriculture (ITPGRFA), the Convention on Crop

Biodiversity (CBD) and the Cartagena Biosafety Protocol of the CBD, affects the movement of germplasm across international boundaries. These factors may influence the relative priority that can be attached to traits and access to suitable genetic sources for introgression into the breeding programme.

SKILLS AND RESOURCES FOR TRAIT DELIVERY. Trait delivery requires the appropriate personnel, skills and resources to be available. For example, if appropriate and representative bioassays for consumer traits are not available, then the new variety design becomes non-viable.

*Cost*

The new variety design must be affordable. The cost of developing a new variety with a broad range of requirements needs careful consideration. This should consider not only whether sufficient finance can be mobilized but also whether the breeding programme is justified, given the scale of the potential use and economic benefits to farmers, consumers and others in the crop value chain. Work planning and cost estimation is required for the introgression of each trait, the total cost of the breeding programme needs to be understood, and a business case needs to be made for the investment (see also Chapter 7, this volume).

*Timeliness of new variety release*

One of the common reasons for the lack of adoption of new varieties, in both developing and industrialized country markets, is a primary focus on the discovery and introgression of individual traits at the expense of optimizing the host germplasm. Consequently, improved traits are introduced, but the new variety does not satisfy key clients. This lack of success in varietal development can be due to a number of reasons. First, there may have been insufficient back checking to ensure that all client/consumer requirements were being met. Secondly, the requirements of key clients may not have been fully understood, or it may not have been possible to address them. Thirdly, time pressures for assessing the performance (often measured by numbers of varieties released) of the breeding programme by research institutions and investors may have led to the premature release of some new varieties before they were able to provide superior performance compared with exiting varieties.

## Creating a Product Profile

The goal for a demand-led breeder is to create a new variety that fulfils a client demand and either improves the design features inherent in existing varieties or provides new benefits that will attract increased varietal adoption. The new design must be easily understood and it must be possible to communicate it to clients, the science delivery team, managers, investors and other stakeholders. This process will help to build ownership and enable iterations to be made to improve the variety design. For clarity, the variety design should fit on to two A4 sheets of paper. The first page is a table of technical attributes with the performance benchmarks to be achieved. The second page is a visualization of all these attributes and of their relative importance in the breeding programme in terms of value creation and market demand.

Syngenta, in partnership with Market Edge, have created a product profiling tool as an Excel-based macro. The profiling tool has been designed to enable a systematic approach to analysing trait identification, ranking and prioritization. It has been kindly provided to the Syngenta Foundation by Syngenta Ag and Market Edge for use in African postgraduate plant breeding programmes and in the continuing professional development of plant breeders in Africa. The product profiling tool is further described in Appendix 4.1 at the end of this chapter, and an electronic copy (e-copy) of the tool is available as Appendix 1 of the e-learning resources for this volume.

Prerequisites for using the product profiling tool are that breeders must have detailed knowledge about the crop under development, its uses and the needs of the clients in their value chain. They must also have the ability to make a series of qualitative and quantitative judgments and assumptions on product performance, trait importance and market demand. These assumptions and the resulting variety design are only as good as the validity of the information used to create the output(s). The four main inputs required for product profiling are:

- **Market research.** Breeders need to commission detailed market research and understand the needs and requirements of all clients in a value chain – from farmers through to consumers.
- **Variety performance.** They need to have detailed technical knowledge of the attributes and performance of registered varieties and of those that are market leaders or are gaining market share.
- **Traits required.** They need to be able to define the list of trait characteristics required and create a performance benchmark for each trait.
- **Trait prioritization.** They need to be able to assess the importance of each trait in terms of its uniqueness/differentiation, its likelihood of being in demand by farmers or by other clients in the value chain, and/or its likelihood of carrying a price premium.

## Stages in product profiling

### *Stage 1: Target market and value chain analysis*

Stage 1 requires breeders to decide for which group of farmers with common needs the variety is being designed. This is the target market segment and requires a name and description (e.g. field-grown tomatoes in the Savannah region of Ghana). The description should include the number of potential users of the variety and indicate where they are based. If there is a market value size already known because of seed sales, this should be included or an estimate provided. The area (hectares) that could be grown with the variety should also be included. The more quantitative and qualitative measures that can define the potential users and the value chain for which the variety is being targeted the better.

It is recognized that some of these values are difficult to estimate. However, their inclusion in the variety description is an essential exercise and it encourages engagement with experts in each parameter and supports connections with seed companies, both as a source of information and as potential partners. A critical component of Stage 1 is identifying the most important clients and the organizations in the value chain that will drive demand, and establishing contact with them. Contact details are required for leading farmers/farmer organizations and other clients in the value chain to enable ongoing consultations to take place during the design of the new variety. A pro forma table for the type of information required in Stage 1 is shown in Table 4.1.

*Stage 2: Trait descriptors of crops*

Stage 2 involves defining the characteristics of landraces and existing varieties and being able to use international nomenclature to describe them. This enables experts and stakeholders to discuss the key characteristics of the current varieties preferred by farmers and all of the possible traits that might be required in the future. Using internationally recognized descriptors has an additional benefit of facilitating easier communication between breeders and holders of potentially useful germplasm containing traits for inclusion in the breeding programme. Demand-led breeders should have a clear set of crop descriptors to create new variety profiles and use these to communicate their breeding goals and objectives.

Trait descriptors are compiled and published by the CGIAR International Agricultural Research Centres and other national and international research institutions and organizations. Examples of trait descriptors for potato and radish are shown in Tables 4.2 and 4.3, respectively. Other example of well-characterized

**Table 4.1.** Product profiling. Stage 1: Target market and value chain analysis. Table showing the key information required for demand-led variety design.

| Client and market information | Description |
|---|---|
| Market segment | |
| Crop use(s) | |
| Country | |
| Region(s) | |
| Agro-ecological zone(s) | |
| Number of farmers | |
| Area of crop grown (000 ha) | |
| Seed market size (000 kg) – current | |
| Seed market size (000 kg) – potential | |
| Seed value (US$ 000) – current | |
| Seed value (US$ 000) – potential | |
| Client contact details for consultations: | |
| Key farmers | |
| Seed companies | |
| Value chain organizations | |
| Government officials | |

**Table 4.2.** Product profiling. Stage 2: Trait descriptors for potato. From Huaman *et al*. (1977).

| Trait group | Trait | Trait description |
|---|---|---|
| Flowering plant | Petal colour | White, white and pink, white and violet, pink, violet |
| | Maturity, time of harvest maturity | Very early, early, medium, late, very late |
| Tuber | Colour of skin | Yellow, red, blue, parti-coloured |
| | Shape | Round to round oval, round oval to long oval, long oval to long, long, very long |
| | Contour of the long and very long tubers | Kidney or pear shaped, irregular |
| | Flesh colour | White, neither clearly white nor clearly yellow, yellow |
| Consumers (home, export, starch industry) | Underwater weight | |
| | Starch (%) | |
| | Crisp quality | |
| | Cooking quality | |

**Table 4.3.** Product profiling. Stage 2: Trait descriptors for radish. From UPOV, 1980.

| Trait group | Trait | Trait description |
|---|---|---|
| Flowering plant | Petal colour | White, white and pink, white and violet, pink, violet |
| | Tuber, tendency to become pitchy | Absent/very weak, weak, medium, strong, very strong |
| | Maturity, time of harvest maturity | Very early, early, medium, late, very late |
| Tuber | Thickness | Thin, medium, thick |
| | Width of root | Thin, medium, thick |
| | Shape of tuber | Transverse/elliptic, circular, elliptic, obviate, broad rectangular, rectangular, narrow rectangular, narrow obtriangular, icicilical [icicle shaped] |
| | Shape of crown | Concave, plane, convex |
| | Shape of base | Acute, obtuse, round, flat |
| | Skin coloration | One coloured, bicoloured |
| | Colour of upper part | White, pink, red, violet |
| | Flesh colour | Translucent, opaque |

descriptors are those of sweet potato (Huamán, 1999), and sorghum (IBPGR/ICRISAT, 1993). Further, the Integrated Breeding Platform (IBP) has documented 'Trait Dictionaries' of ten crop species for breeding (see https://www.integratedbreeding.net).

PLANT BREEDERS' RIGHTS. In practice, a new variety/cultivar has to be registered and should be included on the recommended list of varieties for the country concerned. In order to be on this recommended list, the new variety has to be an improvement over the benchmark variety in reference to a set of variety descriptors set by clients.

Globally, over 73 countries are members of the International Union for the Protection of New Varieties of Plants (UPOV). These countries have agreed to acknowledge breeders' rights resting on a variety. UPOV has issued a set of basic rules for the development of breeders' rights in the countries of its membership (see http://www.upov.int/members/en/).

In order to obtain breeders' rights, a new variety must meet the following (DUSN) requirements, that it is:

- **D** – distinguishable, from all other registered varieties/cultivars;
- **U** – uniform;
- **S** – stable, not changing during maintenance and multiplication; and
- **N** – novel/new, innovative and should not be commercialized already.

These attributes are established through the DUSN trials conducted in each country according to the prescription of the national variety testing agency.

*Stage 3: Product profile*

In the private sector, each variety design usually comprises more than 40 characteristics. It is this profile of features that defines the uniqueness of a new variety, and the range of advantages and differences that it has compared with existing varieties. Systematic analysis of traits is required in each of the six description categories listed below:

- crop yield parameters;
- seed and plant (plant architecture and traits important for seed production);
- biotic stress (fungal, viral, bacterial and nematode pests);
- abiotic stress (environmental traits);
- crop handling parameters (harvest, storage and transport); and
- value chain requirements (consumers, processors and value chain stakeholders).

For each trait: (i) a benchmark variety is required to be included as a quantitative or qualitative performance measure; and (ii) a decision is required on whether the performance of the new variety must be equal to or better than the benchmark variety. In this way, the competitive advantage of each trait is displayed versus the existing varieties that it will be competing with in the market place. In most cases, a performance less than the benchmark is unlikely to be acceptable. An example of an outline product profile is provided below (Fig. 4.1).

*Stage 4: Trait prioritization*

Determining the relative priority of different traits in a new variety design is an important and active process in demand-led variety design. Typically, there are over 40 traits in a new variety design and its product profile. Each trait needs to be assessed for market value in two dimensions, namely, differentiation and market demand (see Fig. 4.2).

- **Differentiation** is measured by willingness to pay a price premium and opportunity to grow a market share. It is a measurement of how different each trait is from what is already available and the benefits that it provides. A top score of ten means that farmers or the value chain recognize the value of

| Trait category | No. | Trait | Trait description | Variety benchmark | Performance required (=, >, >=, >>) |
|---|---|---|---|---|---|
| Crop yield | 1 | | | | |
| | 2 | | | | |
| Plant architecture Seed production | 1 | | | | |
| | 2 | | | | |
| Biotic stress | 1 | | | | |
| | 2 | | | | |
| Abiotic stress | 1 | | | | |
| | 2 | | | | |
| Crop handling, harvest, storage, transport | 1 | | | | |
| | 2 | | | | |
| Value chain clients, consumers, processors | 1 | | | | |
| | 2 | | | | |

**Fig. 4.1.** Stage 3 of product profiling: creating a product profile – an outline product profile.

the trait and are prepared to pay a high price premium; a higher market share is also likely if the price is appropriate. A low score of one means that the trait is very similar to what is already available, no price premium would be accepted and only a low market share could be expected.

- **Market demand** is measured by the percentage of growers (or % total crop area). This measurement relates to the level of farmer demand. A score of ten means that farmers are likely to want to purchase this variety and a high market share is likely. A score of one means that there will be little demand and only relatively few farmers will be interested in purchasing the variety.

All of the traits in these two dimensions are classified and displayed in one of four quadrants, as shown in Fig. 4.2:

- **Essential/must have traits.** These traits must be present in the design for a significant number of farmers to be interested in purchasing and growing the new variety.
- **Winning traits.** These traits are the ones that have a premium value. They will be in demand by many farmers, and the farmers will be prepared to pay a price premium for them. These traits are likely to result in the market value of a variety growing. They deserve serious attention in the breeding programme if they can be delivered at an acceptable cost and within an acceptable time frame.
- **Niche/limited opportunity traits.** These traits offer much added value and are likely to achieve a price premium because of their economic returns, but only to a limited group of farmers (or consumers or other clients). A key question with this group of traits is whether the development costs are justified, given the overall (lower) numbers of farmers and other clients who may be interested in adopting the new variety or paying for its product.

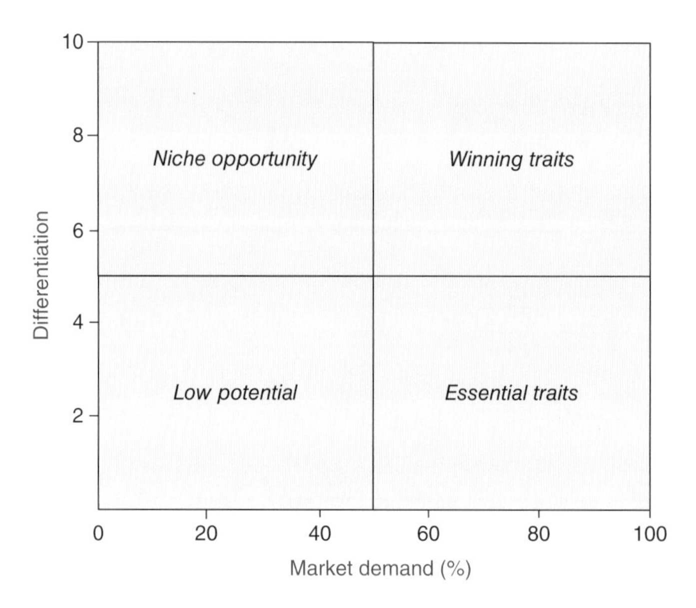


**Fig. 4.2.** Stage 4 of product profiling: trait prioritization as assessed by differentiation (willingness to pay a price premium and opportunity to grow a market share) versus market demand (% growers/area that need the trait). With kind permission from Syngenta and Market Edge.

- **Low potential traits.** These traits provide the lowest value to farmers. They have low differentiation from what is already available in the market, and only low market shares and no/minimal price premiums are expected to be paid.

An example of trait prioritization, based on differentiation and market demand, and leading to four categories of traits, is shown in Fig. 4.3.

### *Stage 5: Trait trade-offs and feasibility*

Once the ranking in importance of each trait has been done and a series of vital traits have been identified, then the feasibility of achieving the full spectrum of required traits together with the level of performance required in a single variety needs to be assessed.

In the event that alterations need to be made to the product profile, revalidation of the new design is required. The design should be discussed with the clients in the value chain and with other stakeholders to ensure that the assumptions that have been made about farmer adoption and demand are still valid before the breeding programme commences.

Further details of the product profiling tool discussed above are contained in Appendix 4.1, Product Profiling Tool, at the end of this chapter. An e-copy of the product profiling tool is available as Appendix 1 of the open-resource e-learning materials for this volume, with grateful acknowledgement to Syngenta and Market Edge for their contribution of this educational resource.

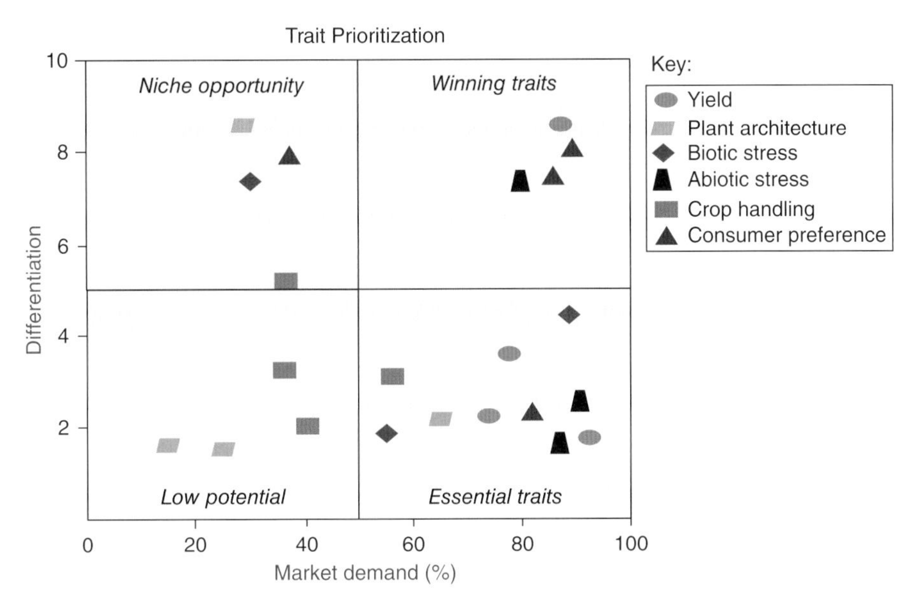


**Fig. 4.3.** Stage 4 of product profiling: visual chart of trait priorities as assessed by differentiation (willingness to pay a price premium and opportunity to grow a market share) versus market demand (% growers/area that need the trait). With kind permission from Syngenta and Market Edge.

## Setting External Performance Standards (Box 4.2)

There is a range of considerations that must be taken into account in a variety design and its performance measures that go beyond the needs and preferences of the value chain. These include the requirements of registration authorities and other government officials and certification schemes; they can also pertain to demonstrating the performance claims that may be made for the new variety. It is critical that these standards are understood at the variety design stage and when breeding goals and objectives are set.

### Variety registration

For agronomic and productivity traits, the requirements of the value chain are usually equal or higher than regulatory requirements and so are not an issue when breeding goals are set. However, registration requirements must be investigated in each target country to be certain. In the case of consumer traits, this is often not the case. Registration in a range of countries in Africa is dependent on demonstrating yield improvements over existing registered varieties. If yield improvement is not included in the new variety, there is serious risk that a highly welcomed consumer trait demanded by the market, such as shelf life or taste, will have registration difficulties in spite of years of investment and endeavour.

**Box 4.2.** Setting external performance standards: educational objectives.

**Purpose:** to ensure that demand-led breeders integrate into their breeding goals and objectives all external performance standards that may influence their new variety release and farmer adoption.

**Educational objectives:**

- to identify and understand the external performance standards that are required to be met for all new demand-led varieties;
- to integrate these performance standards into the variety design and incorporate measures into the breeding goals; and
- to identify potential risks and develop mitigation strategies.

**Key messages**

- It is important to understand the external standards that are required and have clear performance benchmarks for assessing germplasm.
- Clear quantified breeding objectives are essential.
- It is necessary to think broadly across different external requirements from registration, to seed certification, to variety identification and in other areas of expertise, and to seek inputs.

**Key questions**

- Are there variety performance standards you must meet to achieve registration or seed certification for a new variety in your country?
- Which agency sets these standards?
- How can you find out?
- What level of performance is required for each trait?
- How will you measure trait performance?
- Are there harmonized regional standards or are these expected in the future?
- Do you need to create a special kit to be able to demonstrate performance for product promotion, variety identification or the monitoring of farmer adoption (e.g. for a nutrition trait)?

Risk mitigation actions must be taken in advance of a breeding programme starting, such as initiating detailed discussions with government officials about the registration situation and whether exceptions will be granted for consumer-demanded traits. Opportunities need to be sought to understand the attitude of key registration officials and variety release committees to consumer- and demand-led traits, and to find joint solutions. If this results in changes in regulatory legislation being required, the timelines are likely to be lengthy and unpredictable.

When considering the external standards required for registration the following questions need to be addressed:

- Are the characteristics you are seeking already known and part of the official crop descriptors, or are you creating a variety with a very different phenotype (e.g. purple tomatoes, see Botches and Myers, 2009; or pink bananas)?
- What are the minimum requirements set by the national testing agency and national variety releasing committee?

  - Does your design contain traits that adhere to the preferred requirements of farmers or market or industry in a given crop (see examples in Tables 4.2 and 4.3)?
  - Do historic performance data exist (genotype × environment, stability analysis)?
  - Are field and laboratory evaluation protocols available for agencies to make the relevant comparisons?
- What is the level of government interest and urgency for the new variety in the cropping system? Less stringent standard requirements may be allowed for urgently needed varieties (e.g. varieties with resistance to a new disease in the country).
- What is the relative performance of your new variety against existing varieties in the national list of recommended varieties?

### Seed certification

The standards required for mandatory or voluntary seed or plant certification need to be included in the breeding goals.

### Variety identity

Variety identification is an important factor that requires consideration and potential inclusion in setting standards and breeding goals. This is for several reasons. Demand-led varieties that contain consumer traits or other special features such as nutritional qualities need to be easily recognized for authenticity, price differentiation and branding purposes. Buyers will want to know if the variety truly is what it is claimed to be. This can be achieved by including in the design a phenotypic marker/signature so that its authenticity is clear. Alternatively, an identity validation method/diagnostic kit may need to be part of the development plan. Small improvements in nutritional value (or other consumer-preferred traits) may not justify the cost and the effort. However, true innovation and the generation of market-leading varieties may warrant the extra investment in new variety development. The additional benefit of having an identity system means that simpler and more accurate household and market surveys will be possible for tracking varietal adoption.

## Validating New Variety Designs (Box 4.3)

It is clear that creating new varieties with all the ideal trait characteristics is not easy and that trade-offs will often be required. This means prioritizing which of the stakeholders in the value chain have requirements that are more important than others. There is no easy way of achieving this, and careful analysis is required. Farmers and processors should have the greatest role in variety design for staple food crops when eating and cooking qualities are unaffected. This is because in general the new varieties have undifferentiated consumer features and are often bulked and mixed as commodities for uniformity. Here, yield, productivity and production costs are key drivers for design.

**Box 4.3.** Validating new variety designs: educational objectives.

**Purpose:** to ensure that all new variety designs closely reflect the conclusions from market research investigations, and that any revisions to the design that may alter future demand are validated with potential clients before investment in the breeding programme commences.

**Educational objectives:**

- to develop criteria and ways to assess whether or not a new variety design is wanted by farmers and other clients in the value chain; and
- to identify factors that affect farmer variety preference and adoption, and devise a strategy and plan to manage and mitigate the risk factors limiting adoption.

**Key messages**

- It is important to communicate the product profiles and find the best ways to display them and win support for them.
- The visioning of potential landscape changes is essential.
- It is necessary to be clear after trade-off decisions have been made between the ideal profile and its feasibility that the proposed design responds to market demand and will be adopted by farmers.
- The design needs to be validated with potential users.
- There should be a clear plan of engagement with key stakeholders in the value chain during the whole timeline of variety development.

**Key questions**

- How can you be sure that the variety you have designed is what farmers and other clients in the value chain want?

Conversely, the consumer is the prime decider in purchasing higher value fresh fruits and vegetables, especially in urban situations where there is choice, and therefore, consumers' opinions should drive the priority in trait setting for new varieties of these crops.

When the prototype designs that are driving the breeding goals have been decided and significant trade-offs have been made, it is prudent to go back to the key stakeholders in the value chain to show them the revised (new) variety profiles and ensure that there is still demand for these designs. In the case of consumer-based traits, this validation may require further market research and prototype design testing with consumers. As lines are progressed through the stage plan, regular validation of the performance of the potential new varieties with value chain stakeholders is recommended so that they maintain their interest in the new varieties (Anthony, 2013).

## Translating Product Profiles into Breeding Objectives (Box 4.4)

A clear distinction needs to be recognized between designing a target product profile and creating breeding goals and objectives to deliver the new variety. They are not the same thing and demand-led breeders need to effectively convert one to other so that a successful science programme can proceed.

**Box 4.4.** Translating product profiles into breeding goals and objectives: educational objectives.

**Purpose:** to enable participants to create a set of breeding goals and objectives for a new demand-led variety design that are fit for purpose, of high quality and feasible.

**Educational objectives:**

- to be able to translate customer requirements into measurable and achievable breeding goals and objectives; and
- to establish the feasibility of translating client requirements into breeding goals and objectives.

**Key messages**

- Product profiles should drive the setting of breeding goals and objectives.
- The translation of product profiles into breeding goals is an essential part of the design process and iteration of the designs may be required.
- Designs must be achievable with the technology, capability and resources available.
- Performance must be measurable with 'fit-for-purpose' assays.

**Key questions**

- What breeding activities/methods will you follow to achieve the product profile?
- How will you identify and access the germplasm needed?
- How will you combine traits from different sources of germplasm?
- How will you design your breeding strategy to meet the product profile? (e.g. what crosses will be made and when, what population structures will there be?)
- How do you translate a product profile into measurable breeding goals?

This translation has two specific considerations that relate to the feasibility of delivery from both a technical and a practical perspective. New varieties that deliver a step change in performance are highly innovative and require new and often unproven ideas on how to access novel germplasm and achieve unusual combinations of traits. It is common for product profiles to require some revision to increase the probability of delivery. Innovation and ease of delivery can be inversely correlated and paradigm shifts can require supporting investigative science programmes and pre-breeding activities before a full breeding programme can be initiated.

## Feasibility of delivery

### *Scientific feasibility*

A clear step is required to determine how the desired traits can be incorporated together into appropriate germplasm and reach the performance levels required. What is the likelihood of success? Some key questions to consider are:

- What sources of germplasm do you have available with the required traits?
- If not, what options, costs and time will be required to access (or develop) germplasm with the required traits?

- Will you need to do gene discovery and search for new sources of genes, e.g. from wild relatives and a pre-breeding programme?
- Can you construct a viable breeding strategy of crossings and selections?
- What level of increase in performance is required and is there proof of concept?
- Are there fundamental (actual or theoretical) physiological or genetic reasons why particular traits cannot be combined together in the same plant? If so, revisions to the design may be required?

*Practical and legal feasibility*

There are many considerations to gaining access to the genes and traits that a breeder may wish to include in the design. Some of these may be feasible theoretically, but in practice may be intractable and create major risks to the breeding programme. Key questions to consider are:

- If the germplasm is located in another breeding programme, will you be able to gain access?
- Who owns the germplasm or desirable traits?
- Are the ownership rights clear?
- Is there intellectual property involved?
- Will you need to obtain freedom to operate agreements with the germplasm owner and will you be required to make royalty payments on any resulting varieties?
- If the germplasm is located in another country, will phytosanitary and/or other legal regulations prevent you from accessing it?
- Will your request trigger the ITPGRFA and you will be required to complete a standard material transfer agreement and the associated benefits sharing requirements (ITPGRFA, 2015)?
- Is your crop covered by the CBD (2015)?
- Are your home country and/or the host country holding the germplasm parties to these international agreements – ITPGRFA and CBD?

## Performance testing and bioassays

For breeding objectives to be set and achieved, demand-led approaches often require additional bioassays and methodologies to test trait performances. These bioassays are additional to the agronomic- and productivity-based tests that are routine or contained within farmer-participatory trials. Consumer taste and cooking panels are also usually required. These panels may need to be different for urban and rural consumers.

Indicative early-stage high-throughput evaluation tests may also need to be created as part of the pyramid of performance testing for line selections. These tests must be able to provide reliable data for: (i) performance measurement of the desired traits; and (ii) ranking varieties.

During the latter part of the development programme, larger scale testing through commercial equipment will be needed to gain full commitment from value chain stakeholders for the new variety. The results and ranking of varieties from small-scale testing must have a good correlation with larger scale testing. Some questions to consider are:

- Do you have all of the suitable bioassays that are needed to deliver your demand-led breeding goals or are innovative new tests required?
- How confident are you that your bioassays and field trials will adequately predict performance for each trait?
- Will you need to do any new bioassay development?
- Should any new bioassay development be done by your laboratory alone or would it be better to partner with value chain stakeholders in developing joint bioassays?
- Would a public–private partnership or other contractual arrangements be appropriate for bioassay development?

## Challenges in demand-led breeding

Breeding objectives need to be tightly set to deliver the variety specifications. Three frequently encountered problems that lead to suboptimal breeding programmes are described below.

*Inadequately defined product profile*
This may result from factors such as: (i) incomplete discussions with the value chain clients; (ii) lack of detailed knowledge about the performance of frequently used varieties; and/or (iii) lack of definition of the level of performance that is required to create demand. The required traits are often known, but standard quantitative benchmarks for all of the desired traits are not set.

*Poorly defined breeding goals*
Breeding goals may be vague, aspirational and not adequately underpinned with a set of specific breeding objectives to drive the science programme. For example, a broadly defined breeding goal, such as 'to increase drought resistance in sorghum in Africa', needs to be supplemented with more specific breeding objectives that contain clear quantitative measures to reach the goal. In the sorghum example, the specific objectives to drive a demand-led science programme for farmers and their value chain clients could be: 'To increase grain yield as measured by fresh grain weight (kg/ha) by 10%, compared with the best registered varieties when 25% below average rainfall occurs between $x$ and $y$ weeks after germination, and the plant is between growth stages $y$ and $z$, while retaining grain colour, taste, grain milling and cooking time for the most popular varieties used for food'.

*Performance standards either not set or not met*
Demand-led breeding requires a performance standard to be set that is not related to achieving a variety registration but is set to serve farmers and the value chain to create demand. Conventional breeding follows a philosophy of best endeavours and incremental gain through selection. Lines are progressed that

represent the best achieved in a particular year. In demand-led breeding, line progression only occurs if the annual breeding target hurdles are achieved. This is an important difference in philosophy between the usual breeding practices and demand-led breeding. Meeting quantitative targets is given greater emphasis in demand-led breeding progression decisions and, in practice, the ideal benchmark is not always found; here, understanding the range of acceptable quantified metrics enables good decision making by breeders and their project teams.

Rather than registering varieties with small enhancements that ultimately may not attract widespread support and use, demand-led design requires the breeding team to adhere to the agreed breeding goals and targets set for line progression, with confidence that this rigorous standard setting will result in increased farmer adoption at commercialization.

Clearly, novel features may be discovered that may modify the target variety profile. However, the challenge is not to reduce the client requirements for line progression when technical difficulties arrive. Variation from the product profile should only occur in cases where new information or market research indicates that this change will not be detrimental to variety adoption and market acceptance (see also the section on **Validating New Variety Design** earlier in this chapter).

## Learning Methods (Box 4.5)

Before this chapter concludes, a summary is provided in Box 4.5 of learning methods – together with assignments and assessment methods – for use with the main topics that have been covered in the chapter: New Variety Design and Product Profiling; Setting External Standards; Validation of New Variety Designs; and Breeding Goals and Objectives.

**Box 4.5.** Learning methods, assignments and assessment methods.

**New Variety Design and Product Profiling**

*Learning method*

- Presentation on the core principles of product profiling and creating new variety designs.
- Exemplification with a crop and detailed product profile for all key characteristics required by all clients in a value chain.
- Participants (either in groups or as individuals) use the product profiling tool to create new variety designs using parameters provided by the course lecturer.
- Group discussion on product profiles of different crops, comparing and contrasting important aspects for different clients in the value chain.
- Individual or group discussion on sorghum case study (Timu *et al.*, 2014), analysing and comparing the importance of yield, processing and consumer traits of sorghum and their effect on farmer adoption.

*Continued*

**Box 4.5.** Continued.

- Decision-tree analysis: course lecturer to provide a set of parameters for participants to review and make decisions on trait trade-offs to finalize and validate a product variety design.

*Assignment*

- Investigate, compare and contrast the characteristics/descriptors of two lead varieties of a target crop in your home market, and create a product profile for each variety.
- Create an ideal product profile for a new variety with benchmarking versus existing varieties and the plant breeding action plan needed.

*Assessment*

- Assignment on the characteristics of leading varieties and designing a new variety with a prioritized set of traits for starting a breeding programme for a new variety design (as above).
- Exam questions on the core principles and activities needed to create a new variety product profile.

**Setting External Standards**

*Learning method*

- Group discussions on variety registration and the external standards set in participants' countries and the advantages, disadvantages and issues of variety identification markers and monitoring tests.

*Case studies*

- Find an example that demonstrates the issue of registering demand-led traits versus yield performance requirements (e.g. early-season potatoes in Kenya that can achieve a price premium).

*Assignment*

- Summarize all required registration performance standards traits for a target crop in your home country and in neighbouring countries.
- Consult national crop breeding strategy documents to see whether the new variety designs will contribute to government targets.

*Assessment*

- Presentation of assignment.

**Validation of New Variety Designs**

*Learning method*

- Presentation of concepts by course lecturer.
- Group discussion on the relative importance of the different opinions of stakeholders in the value chain for different crop variety designs.

*Assignment*

- To identify who the stakeholders are for each trait in the product profile, assign a level of importance to each stakeholder's views and decide when consultations on the design and breeding progress should take place in the stage plan. Define when and how these interactions might take place.

*Continued*

**Box 4.5.** Continued.

*Assessment*

- Assignment.

**Breeding Goals and Objectives**

*Learning method*

- Presentation on principles of translating a product profile into breeding goals and objectives.
- Participants to review a plant breeding project proposal submitted to an international donor or documented information from a current breeding programme, and have a group discussion on: (i) the relative emphasis and balance between technology-led versus demand-led principles being proposed; and (ii) the quality of the breeding goal and objectives, and whether they should be strengthened to be more demand driven.
- Create a breeding goal and set of breeding objectives for a research proposal based on a demand-led new variety design/product profile.

*Assignment*

- Create a breeding goal and a set of breeding objectives for a new demand-led variety design/product profile to be delivered in your breeding programme. Identify the bioassays required for performance measurement and where they will fit into your performance testing plan as part of the stage plan. Determine what engagement is required with the value chain stakeholders to deliver the testing plan.

*Assessment*

- Assignment.
- Exam questions on the core principles of translating a demand-led product profile into a set of breeding goals and objectives.
- Ability to create a breeding goal and set of breeding objectives from a product profile provided in the exam questions.

## Conclusion

As evidenced in the large private sector seed industry, when demand-driven product design is introduced into productive plant breeding programmes, and combined with excellent science and technology, development rigour, and appropriate awareness campaigns with farmers and customers, significant gains in adoption rates and market share can be made (Anthony, 2013).

The challenge is finding a cost-effective way to integrate and tailor these demand-led approaches to new variety design, product profiling and measurable success criteria so that they fit into public sector and small seed company plant breeding programmes in developing countries – which operate with considerably fewer resources than those of more developed countries. Specifically, ways need to be found to harness the skills and cooperation of the private sector and to better understand tropical crop value chains by working with public and private sector experts to share knowledge and solve problems together.

## Resource Materials

The open-resource e-learning materials available for Chapter 4 include: (i) Appendix 1, an electronic copy (e-copy) of the Excel-based macro product profiling tool developed by Syngenta and Market Edge which is further described in Appendix 4.1 of this chapter; and (ii) a set of slides available for this chapter as part of Appendix 3 that summarize the chapter contents and provide further information. The e-learning material is available at http://www.cabi.org/open-resources/93814 and also on a USB stick that is included with this volume.

## References


Anthony, V. (2013) Demand-driven variety design. Commissioned article from USAID/Syngenta Foundation for Sustainable Agriculture. Available at: http://media.wix.com/ugd/ad2c36_49f5003c432d2d6a4281626ccca63ea6.pdf (accessed 4 May 2017).

Bai, Y. and Lindhout, P. (2007) Domestication and breeding of tomatoes: what have we gained and what can we gain in the future? *Annals of Botany* 100, 1085–1094.

Boches, P. and Myers, J. (2009) The Purple Tomato FAQ. Department of Horticulture, Oregon State University, Corvallis, Oregon. Available at: http://horticulture.oregonstate.edu/purple_tomato_faq (accessed 4 May 2017).

CBD (2015) Convention on Biological Diversity. Available at: http://www.cbd.int/ (accessed 4 May 2017).

Huamán, Z. (ed.) (1999) Section 1.2. Botany, origin, evolution and biodiversity of the sweet potato. In: *Sweet Potato Germplasm Management (*Ipomoea batatas*): Training Manual*. International Potato Centre (CIP), Lima, Peru. Available at: http://1srw4m1ahzc2feqoq2gwbbhk.wpengine.netdna-cdn.com/wp-content/uploads/2016/04/Sweetpotato-Germplasm-Management-Ipomoea-batatas-Training-Manual.pdf (accessed 4 May 2017).

Huaman, Z., Williams, J.T., Salhuan, W. and Vincent, L. (1977) *Descriptors for the Cultivated Potato*. International Board for Plant Genetic Resources, Rome. Available at: http://www.bioversityinternational.org/uploads/tx_news/Descriptors_for_the_cultivated_potato_381.pdf (accessed 4 May 2017).

IBPGR/ICRISAT (1993) *Descriptors for Sorghum (*Sorghum bicolor *(L) Moench)*. International Board for Plant Genetic Resources (IBPGR), Rome and International Crop Research Institute for the Semi-Arid Tropics (ICRISAT), Patancheru, India. Available at: http://eprints.icrisat.ac.in/8639/1/RP_08761_Descriptors_for_sorghum.pdf (accessed 4 May 2017).

ITPGRFA (2015) International Treaty on Plant Genetic Resources for Food and Agriculture. Food and Agriculture Organization of the United Nations, Rome. Available at: http://www.fao.org/3/a-i0510e.pdf (accessed 4 May 2017).

Timu, A.G., Mulwa, R.M, Okella, J. and Kamau, M. (2014) The role of varietal attributes on adoption of improved seed varieties: the case of sorghum in Kenya. *Agriculture and Food Security* 3: 9. Available at: https://agricultureandfoodsecurity.biomedcentral.com/articles/10.1186/2048-7010-3-9 (accessed 4 May 2017).

UPOV (1980) *Guidelines for the Conduct of Tests for Distinctness, Homogeneity, and Stability (Radish)*. International Union for the Protection of New Varieties of Plants (UPOV), Geneva, Switzerland. [Later (1999) version available at http://www.upov.int/en/publications/tg-rom/tg064/tg_64_6.pdf. (accessed 4 May 2017).]

Wikipedia (2015) Market segmentation. Available at: http://en.wikipedia.org/wiki/Market_segmentation (accessed 4 May 2017).

## Resource Materials, including Web Resources

### Plant production guidelines

Burger, E. and Kilian, W. (2014) *Guidelines for the Production of Small Grains in the Summer Rainfall Region*. ARC-Small Grain Institute, Bethlehem, University of the Free State, Bloemfontein and SAB Maltings (Pty) Ltd, Sandton, South Africa. Available at: http://www.arc.agric.za/arc-sgi/Product%20Catalogue%20Library/Guideline%20for%20production%20of%20small%20grains%20in%20the%20summer%20rainfall%20area.pdf (accessed 4 May 2017).

Directorate Plant Production (2010) *Dry Beans: Production Guideline*. Department of Agriculture, Forestry and Fisheries, Pretoria. Available at: www.nda.agric.za/docs/Brochures/prodGuide DryBeans.pdf (accessed 4 May 2017).

Directorate Plant Production (2010) *Production Guideline for Wheat*. Department of Agriculture, Forestry and Fisheries, Pretoria. Available at: http://www.daff.gov.za/Daffweb3/Portals/0/Brochures%20and%20Production%20guidelines/Wheat%20-%20Production%20Guideline.pdf (accessed 4 May 2017).

### International conventions and treaties

Convention on Biological Diversity (CBD). Details available at: http://www.cbd.int/ (accessed 4 May 2017).

International Treaty on Plant Genetic Resources for Food and Agriculture (ITPGRFA). Details available at: http://www.planttreaty.org/ (accessed 4 May 2017).

## Appendix 4.1: Introduction to Product Profiling Tool for New Variety Design

An e-copy of this product profiling tool is available as Appendix 1 of the open-resource e-learning material for this volume. The e-learning material is available at http://www.cabi.org/openresources/93814 and also on a USB stick that is included with this volume.

### Purpose

The product profiling tool (Appendix 1) is an Excel-based macro that has been developed to aid professional breeders to design new varieties. It has been kindly provided here by Syngenta and Market Edge as an educational tool for use in the postgraduate plant breeding programmes of African universities and in continuing professional development programmes for plant breeders in Africa.

Specifically, the e-learning tool provides the following elements:

- **A systematic approach.** This defines the traits required in a variety for a crop segment, and their relative importance and priority.

- **A decision and conclusions framework.** This provides a methodology for individuals and groups of breeders, scientific experts and representatives of a crop value chain to discuss, reach a common understanding and decide the priorities for a breeding programme. It also provides a high-quality, consistent methodology for institutional managers and other stakeholders, including investors, to evaluate and monitor new variety design.
- **A visual communication graphic.** The graphic shows the full set of traits and can be used as a simple diagram to communicate the relative importance of a large number of potential traits to be incorporated into a breeding programme. It can be used also to capture and convey the conclusions from the discussion and decisions taken on priority traits to be included in a new variety.

The profiling tool comprises six individual data sheets (Sheets 1–6) that are linked and which require expert inputs about key clients, the value of markets, plant descriptors and the results of group discussions. These enable the gathering and finalizing of opinions on the importance and relative priority of many different types of traits. Sheet 7 is a summary of conclusions and the actions required. Sheet 8 contains explanatory notes on how to complete Sheets 1–7.

## Product profiling tool: Sheets 1–8

### *Sheet 1. Client and market segment definition*

The first sheet summarizes the target market segment, number of clients, their location and the market value of the crop (within the target market segment).

### *Sheet 2. Crop descriptors*

This sheet needs to be completed using the internationally recognized crop descriptors so that all traits follow the same nomenclature.

### *Sheet 3. Product profile: traits and benchmark varieties*

This is a template to complete with a *de novo* list of traits needed in the new variety design. Each trait is placed in the following categories:

- crop yield parameters;
- seed and plant (plant architecture and traits important for seed production);
- biotic stress (fungal, viral, bacteria and nematode pests);
- abiotic stress (environmental traits);
- crop handling parameters (harvest, storage and transport); and
- value chain requirements (consumers, processors and value chain stakeholders).

There is space on the template for a maximum of 15 traits per category, i.e. a total of 75 traits for each design. For each trait, a benchmark variety is required to be included as a quantitative or qualitative performance measure. A decision is required for each trait on whether the performance of the new variety must be equal to or better than the benchmark. In most cases, performance less than the benchmark is unlikely to be acceptable.

*Sheet 4. Profile input*

The titles of the traits that were put into Sheet 3 are automatically transferred by the macro on to Sheet 4. Each of the six categories of traits is visible by scrolling down the choices within 'SELECT CATEGORY TO INPUT'. To estimate demand, an assessment is needed that rates each trait in two ways (see Indexes 1 and 2 below) on a scale of 1–10, where 10 = high and 1 = low.

INDEX 1: DIFFERENTIATION. A top score of 10 means that farmers or the value chain recognize the value of the trait and are prepared to pay a high price premium (a high market share is also likely if priced appropriately). A low score of 1 means that the trait is very similar to what is already available, no price premium would be accepted and only a low market share is expected.

INDEX 2: IMPORTANCE TO FARMERS. This measurement relates to the level of farmer demand. A score of 10 means that farmers are likely to want to purchase this variety and a high market share is likely. A score of 1 means there will be little demand and only a relatively few farmers will be interested in purchasing the new variety.

*Sheet 5. Product profile: trait visualization graph*

A key benefit of this tool is that a single graph is automatically plotted that provides visualization and categorization of all of the traits into the four quadrants, based on the scores provided for trait differentiation and importance for farmers. This allows groups of breeders and scientists to discuss and agree on the relative importance of different traits, as a part of agreeing on the new variety design and setting the breeding goals and objectives (see also Fig. 4.3).

*Sheet 6. Product profile: implications*

This sheet summarizes the conclusions of the breeder and his/her project team about the relative importance of all of the traits, and which three (or more) traits should be central in the product profile and be the primary focus of the breeding programme. Specifically, the actions needed to minimize risks and maximize opportunities are requested, and a double check is made on the validity of the data sources used to form the conclusions.

*Sheet 7. Conclusions*

Sheet 7 captures the conclusions and actions that are needed to progress the plant breeding programme to include each trait. These actions may be technical, financial, resource-based manpower or other actions that are needed to proceed.

*Sheet 8. Explanatory notes*

This sheet provides the user with guidance from Syngenta and Market Edge on how to use the product profiling tool and the information needed to complete Sheets 1–7.

# 5 Variety Development Strategy and Stage Plan

ROWLAND CHIRWA*

*CIAT (International Center for Tropical Agriculture), Chitedze Agricultural Research Station, Lilongwe, Malawi*

## Executive Summary and Key Messages

### Objectives

**1.** To enable plant breeders to construct a high-quality and well-documented demand-led variety development strategy and a development stage plan in order to enable good governance, rigorous decision making and activity planning within a demand-led breeding project.
**2.** To ensure that the variety development strategy and stage plan allow for the involvement of stakeholders at key decision points, specifically in: (i) the design, development and release of new varieties; (ii) enabling new varieties to reach farmers; and (iii) providing feedback on product performance and farmer adoption.

The chapter contains five main sections. The first, **New Variety Development Strategy**, discusses the creation of a *de novo* demand-led new variety development strategy as a strategic communications document. The second, **Development Stage Plan**, aims to present a clear understanding of the key components and benefits of a demand-led development stage plan that contains the key decision points and information needed for line progression. The stage plan includes all of the required activities and timelines for the creation of new varieties of the target crop. The third section, **Timelines and Critical Paths**, underlines the value of organizing the activities involved in developing demand-led varieties into an optimized plan and describes how to determine the critical path sequence and conduct critical path analysis. The fourth section, **Risk Management**, discusses

*E-mail address: r.chirwa@cgiar.org

the implementation of risk mitigation measures to reduce the likelihood of delays and ensure that outputs are delivered on time. Lastly, the section on **Variety Registration** aims to promote an understanding of the requirements and timescale needed to register a new improved variety for a crop in a country or region. This includes an ability to be able to engage with variety registration officials to ensure that registration procedures are tailored for additional market-demanded traits and to seek their support during the varietal development process.

## How does demand-led variety development add value to current practices?

- **Development strategy.** Demand-led breeding takes an integrated approach to new variety development. It requires a comprehensive analysis of the following: Who are the targeted clients? What are their needs and how may these change? What are the technical and regulatory elements of plant breeding? How will the new variety reach the targeted clients and will their requirements be satisfied? Success is measured by satisfying the demand encapsulated in the product profile and by feedback from farmers on product performance and variety adoption. This requires a comprehensive strategy that creates the framework for: the delivery of new variety design and variety creation; registration and release; client awareness building; seed distribution to farmers; and performance and adoption monitoring.
- **Development stage plan.** Demand-led breeding requires a stage plan to be created with transparent time points and timelines for data review and germplasm progression decisions that involves the participation of key clients in the value chain. This helps to maintain the commitment of clients to new designs, enables joint problem solving, manages expectations and stimulates demand.
- **Development planning.** Demand-led breeding creates more complexity as a result of a broader range of client involvement, trait targets and performance testing. Therefore, to counteract potential delays, greater emphasis is placed on breeders developing professional planning skills, the understanding of critical paths and risk mitigation strategies.
- **Participatory breeding.** Demand-led breeding includes, but goes beyond, farmer-participatory breeding. It puts more emphasis on regularly consulting and understanding the needs and preferences of all clients and stakeholders in a crop value chain. It involves seeking information from farmers and consumers, in both rural and urban areas, through participatory appraisal methods. Consultation is a continuous requirement throughout the whole development process, registration and launch, so that a new variety not only supports farmers' requirements for crop productivity and sufficient food for home consumption, but also ensures that production surpluses can enter markets. A development stage plan that includes the joint development of ideas and joint decision making with

stakeholders in the value chain is critical for success (see Chapters 3 and 4, this volume).

- **Variety design and benchmarking.** Demand-led breeding places emphasis on the systematic, quantitative assessment of varietal characteristics and creating product profiles with benchmarks for varietal performance and line progression. Consumer-demanded traits are given more importance. It requires the prioritization of the many traits desired by farmers, processors, seed distributors, transporters, retailers and consumers (see Chapter 4, this volume).
- **Registration standards.** Early contact with registration officials is required at the variety design phase, well before a potential new variety is ready to enter official registration trials. Thus, at an early stage, there is a need to validate designs, agree standards for consumer-based traits, and create interest in the new variety by officials, as this will potentially accelerate the timelines to delivery of the demand-led varieties (see Chapter 4, this volume).

*Implications for the role of the plant breeder*

- **Market and business knowledge.** To be effective at consultation within the value chain, demand-led breeders require greater knowledge about crop uses, markets and the 'business of plant breeding', which takes the best practices from demand-led breeding and integrates these with the best practices in business.
- **Decision making and planning skills.** Breeders need to further develop their planning skills to deliver two essential aspects of demand-led breeding: (i) a stage plan that is used to plan decision points; and (ii) an activity delivery plan that sits within and supports the stage plan, and contains all the activities necessary to deliver a new variety from initial design through to use by farmers. The different stages are directly linked, so they must communicate with each other.
- **Liaison with government officials.** Breeders need to work closely with key government officials to sensitize them to the need for new demand led-varieties which may have new consumer-preferred traits that respond to market demands; by so doing, they will solicit the support of such officials for new demand-led varieties and advocate the use of the new varieties (see Chapter 4, this volume).
- **Data-driven decisions.** They need to use data that are important for the value chain to make decisions, and to find creative ways to shorten timescales for variety development by using critical path analysis and risk mitigation while in the demand-led variety development process.
- **Compelling business cases.** Breeders need to be able to understand who their clients are and create varieties that have benefits for all in the value chain, as well as delivering an attractive return on investment. They also need to develop a broader and deeper understanding of the range of costs required to develop demand-led varieties as a basis for creating business investment cases that are persuasive to government officials, private and public investors and other stakeholders, so that

they can secure and retain support for a demand-led breeding programme (see Chapter 7, this volume).

## Key messages for plant breeders

*Development strategy*

- **Development strategy.** A comprehensive, delivery-orientated development strategy is essential for a successful demand-led breeding project/programme; this strategy, as agreed with clients and stakeholders, is also an important communications document.
- **Measuring success.** The strategy should contain key performance indicators (KPIs), as part of its monitoring, evaluation and learning plan. Success is measured by variety adoption by farmers; variety performance versus expectations in the variety profile and feedback from farmers and other clients in the value chain (see Chapter 6, this volume).

*Development stage plan*

- **Stage plan.** A development stage plan is essential for every successful demand-led breeding programme. This makes transparent the key decision points, the information required and the involvement of clients along the value chain at the key decision points. A clear plan of engagement with stakeholders is required in the value chain during the whole timeline of variety development.
- **Credibility.** An optimized stage plan can give credibility to a breeding programme, speed up variety development and decision making and, by the involvement of clients, maintain their interest and create demand for the new variety during its development.

*Timelines and critical paths*

- **Timeline optimization.** Demand-led approaches involve more data generation and greater complexity of interactions and consultations with clients. These factors increase the potential for costs and delays. To counteract this, demand-led breeding requires a stronger focus on project planning and the inclusion of bioassays and performance testing, critical path analysis and risk management.

*Variety registration*

- **Early engagement with officials.** Breeders should work closely with government officials to sensitize them to the need for new varieties to include new traits that respond to market demands, and seek their support for new demand-led varieties. Early discussions are needed during the design phase if new consumer traits are likely to require special procedures or are not recognized under existing registration and release procedures (see Chapter 4, this volume).
- **Registration strategy.** Harmonized variety registration requirements in different countries/regions should be explored so as to enable the broadest base of clients to have access to new demand-led varieties.

### Additional key messages for research and development (R&D) leaders, managers, government officials and investors

*Development strategy*

- **Development strategy.** A clear, reviewed, owned and approved strategy is required that provides not only a framework for the design and creation of new varieties, but also awareness of and access to new seeds by farmers, and the tracking of variety adoption and performance versus the targets.
- **Incentives.** Managers should support and encourage their breeders to find the best ways of maximizing reach and scaling seed availability so that seeds reach farmers through both public and private sector seed systems.

*Timelines and critical paths*

- **Timeline optimization.** Managers need to support their breeders to develop optimized development plans, reduce complexity where possible and, where appropriate, partner with the private sector for access to specialized bioassays and as a means of reducing costs.

*Variety registration*

- R&D managers should facilitate and support their breeders in consultations with registration officials and enable forward-thinking about registration requirements for consumer traits.

## Introduction

The objectives of this chapter are:

**1.** To enable plant breeders to construct a high-quality and well-documented demand-led variety development strategy and a development stage plan in order to enable good governance, rigorous decision making and activity planning within a demand-led breeding project.
**2.** To ensure that the variety development strategy and stage plan allow for the involvement of stakeholders at key decision points, specifically in: (i) the design, development and release of new varieties; (ii) enabling new varieties to reach farmers; and (iii) providing feedback on product performance and farmer adoption.

The chapter contains five main sections. The first, **New Variety Development Strategy**, discusses the creation of a *de novo* demand-led new variety development strategy as a strategic communications document. The second, **Development Stage Plan**, aims to present a clear understanding of the key components and benefits of a demand-led development stage plan that contains the key decision points and information needed for line progression. The stage plan includes all of the required activities and timelines for the creation of new varieties of the target crop. The third section, **Timelines and Critical Paths**, underlines the value of organizing the activities involved in developing demand-led varieties into an optimized plan and describes how to determine the critical path sequence and conduct critical path analysis. The fourth section, **Risk Management**, discusses the implementation of risk mitigation measures to reduce the likelihood of delays and ensure that outputs are delivered on time. Lastly, the section on **Variety Registration** aims to promote an understanding of the requirements and timescale needed to register a new improved variety for a crop in a country or region.

The aim of the chapter is to enable African plant breeders to understand the principles and share the best practices in the construction of a variety development strategy and stage plan for demand-led variety design, and to act as a resource for education in this field. For this purpose, boxes are included in several sections of the chapter that summarize their educational objectives and present the key messages and questions that are involved. There is also a final box at the end of the chapter that summarizes the overall learning objectives.

### Key principles of and differences between creating a development strategy and a development stage plan

*Demand-led development strategy*

- The strategy defines the core characteristics of the new variety product, its competitive advantages and the key goals to be achieved.
- It requires a full analysis and understanding of the external environment and of all of the influencing drivers, current and foreseen. It focuses primarily on external factors and what choices a breeder will make to serve clients.
- It considers key questions on 'What?', 'Why?' and 'For whom?'
- The strategy has a broader scope than a plan and considers the end product and its adoption.
- Strategy creation should precede planning. The stage plan should be created to deliver the new variety and its strategic goals.

*Development stage plan*

- The development stage plan is a series of steps for the design, creation and delivery of the new variety to farmers. Planning involves looking at all possible paths to achieve this goal.
- It involves making decisions on how to use resources and the actions needed to achieve the development strategy.
- The stage plan requires consideration of questions involving 'How?', 'When?', 'Where?' and 'Who?'

## New Variety Development Strategy (Boxes 5.1 and 5.2)

A central and differentiating principle of conducting demand-led breeding is its holistic and client-driven approach to new variety design and development.

**Box 5.1.** Elements of a new variety development strategy.

**Target crop agricultural landscape**

- Current crop supply and demand: domestic and import data.
- Key challenges.
- Policy landscape and targets.
- An enabling environment.

**Market analysis**

- Crop uses.
- Market segments.
- Farmers.
- The value chain.
- The seed production and seed retail system.

**Target clients and market segment**

- Quantification of market demand.
- Clients targeted – numbers of farmers, client profiles, farm sizes, farm systems, countries, geographic scope, agro-ecological zones.
- Crop segment and use.
- Varieties used – profile, history, performance, strengths and weaknesses.

**Variety design and market positioning**

- New variety product profile.
- Variety design attributes and quantitative benchmarks.
- Differentiation from existing varieties.
- Means of identifying the variety after registration and post-launch monitoring of adoption.
- Ranking of trait priorities.
- Breeding goals and objectives.

**Development plan and timetable**

- Stage plan milestones and decisions.
- Operational activities plan and timetable.
- Critical path.
- Risk mitigation actions.
- Skills and expertise required.

**Development costs**

- Quantified costs of the new variety development programme.

**Development investment case: benefits and costs**

- Quantitative and qualitative benefits – social, economic, trade, environmental, etc. and how this case fits within national priorities, compared with its costs and risks.

**Project governance**

- How decisions will be taken on line progression through the stage plan.
- The involvement of key stakeholders.

*Continued*

**Box 5.1.** Continued.

**Variety demand: awareness, promotion and scaling**

- Communication and advertising.
- The involvement of extension services, lead farmers and the value chain.
- Farmer demonstrations.
- The seed production plan.

**Seed systems**

- How will the variety reach the target farmers?
- What will the distribution system be?
- What partnerships are required?

**Variety performance and adoption**

- Who will be responsible for the post-launch tracking plan and how will it be done?
- What is the client feedback on variety performance?

**Performance measures and risk management**

- Demand-led performance measures.
- Clear identification of what success looks like – target setting.
- Monitoring and evaluation and learning plan.
- Key delivery risks and risk mitigation plan.

This involves an investigation, at the early ideas stage, of the key information needed to define and deliver varieties for end users and their market value chains. It also involves seeking buy-in and support from clients, key stakeholders and influencers such as government officials, who affect supply and demand. To instil clarity and rigour, a demand-led development strategy should be created for each new variety design. Each new variety will have its own strategy document that will need to be reviewed periodically as the breeding programme progresses. When several new varieties are developed for the same clients, it may be appropriate to include these within the same strategy document.

The creation of a product development strategy requires an understanding of the following factors: the strategic fit of the product(s) with national government crop priorities; a systematic review of the crop supply and demand situation; definition of the clients being targeted, their needs and preferences; clarity about the clients' market segments and value chains; definition of the problem being addressed; the variety design profile; current variety performance; the product development plan; development costs; project governance; the involvement of stakeholders; risk management measures; performance measures; and feasibility. The outputs from consideration of these factors are used to evaluate the return on investment and, if they are of high enough priority, breeding goals and objectives are set to deliver the new varieties and monitor their use.

**Box 5.2.** Demand-led new variety development strategy: educational objectives.

**Purpose:** to be able to create and communicate a development strategy for each new variety design.

**Educational objectives:**

- to understand the core components of a new variety development strategy; and
- to analyse market information and combine this with technical and scientific knowledge to create a development strategy for a new variety.

**Key messages**

- A clear demand-led development strategy and delivery plan should be created for each new variety or each group of varieties to be developed.
- The development strategy document should contain the stage and activity plan and be used as the primary communications tool to gain input and buy-in from managers, variety development team(s), stakeholders and investors.
- A development strategy should be created before the breeding programme commences.

**Key questions**

A demand-led variety development strategy should address the following questions:

- Who is the new variety for?
- How many farmers are there who are involved and where are they?
- What markets and market segments do these farmers serve?
- What is their value chain?
- What is the crop used for?
- What is the demand and how may it change in the future?
- What are the unique characteristics of the variety?
- What is the benefits case for clients to grow this new variety to serve their markets?
- What is the investment case for development (including how it fits within government national priorities)?
- Why should financial investment be provided?
- Who will develop the variety and how will clients and their value chains be involved?
- How will the development be governed?
- What is the variety development and registration plan?
- What seed system will be used to supply clients with the new variety?
- How will adoption by clients be monitored?
- What are the critical success factors and performance measures?
- What is the feasibility and what risk management is required?

An enabling environment is required for engagement with the value chain and private sector actors so that more modern varieties will reach smallholder farmers. Demand-led approaches mean that these aspects have been considered and possible partnerships explored before breeding work starts. The seed channel and supporting legal frameworks that will encourage private sector engagement require special consideration, as they can make or break access to germplasm and the scale at which seed can be delivered to farmers. Demand-led breeders are not expected to create new seed systems for their varieties, but they do need to engage with the relevant professionals to establish how seed and product information will reach farmers. This involves liaising with extension professionals, government officials and the private sector, where public and private seed systems are operational. In other situations, it may be necessary to stimulate action by liaison with key individuals in their subregional organizations (SROs), foundations, non-governmental organizations (NGOs) or organizations such as the Alliance for a Green Revolution in Africa (AGRA), who have a mission to facilitate and develop seed systems for farmers to access improved seed.

### Creating a demand-led development strategy (Box 5.1)

The framework for creating a new variety development strategy is shown in Fig. 5.1 and the key elements of the strategy are shown in Box 5.1. These elements are covered in detail in the preceding chapters of this volume. Best practice involves condensing the information obtained on all of these issues into a new variety development strategy document of no more than ten pages in length. This strategy document then becomes a primary communications tool for discussion with management, the multifunctional development team, and the stakeholders and investors, in seeking their input, providing information and gaining ownership and approval for the new variety development.

Box 5.2 outlines the educational objectives that are involved in creating and designing a development strategy for each new variety design.

## Development Stage Plan (Box 5.3)

The stage plan is the tool that a demand-led breeder should use to govern the decision making and progression of leading lines towards commercialization.

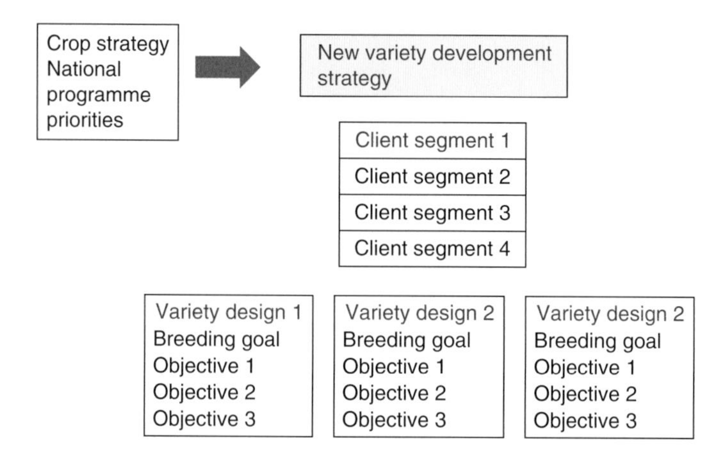


**Fig. 5.1.** Framework for a new variety development strategy. Provided by V.M. Anthony.

**Box 5.3.** Development stage plan: educational objectives.

**Purpose:** to be able to create a development stage plan to deliver each new variety design.

**Educational objectives:**

- to understand the key principles, components and benefits of using a development stage plan to deliver a successful demand-led breeding programme; and
- to develop a detailed development stage plan that includes all of the required decision points, and the activities, data collection and analysis required to make these decisions, together with the associated timelines and costs, to create and register a new variety of the crop of interest.

**Key messages**

- A development stage plan is essential for all breeding programmes.
- A stage plan is the key tool for supporting timely decision making and determining whether a new genotype/line is ready for progression to the next stage and further investment.
- A stage plan contains all of the activities needed to deliver a new variety, from initial design to use by farmers. The activities are mapped on to all of the different stages of progression along the timeline. The different stages are directly linked and so they must communicate with each other.
- A stage plan is especially useful not just to organize activities but also to gain inputs, support and pull through from clients at the right time in the development process.
- A common language is especially important for effective communication between teams of clients, and with experts from different disciplines and across each stage.

**Key questions**

The key questions to be addressed include:

- What is a demand-led development stage plan?
- What are the benefits of creating a stage plan and how can it contribute to the success of a breeding programme?
- How can a stage plan encourage demand?
- How can a stage plan be used to support quality decisions and ownership of these decisions by stakeholders?
- What and when are the critical decision points on varietal design during the stage plan?
- Who should you involve to make a progression decision from one stage to the next?
- What is the full list of activities and information/data required to progress from one stage to the next?
- What expertise and skill is needed for each activity?
- Which activities can be done concurrently and which must be done sequentially?
- How and when will you do critical bioassays/genetic evaluations to test for the presence of essential consumer traits, agronomic traits, processing traits and production traits?
- What are the resource requirements and costs of each activity and each stage?
- How easy is it to procure these resource needs?

It is a transparent framework that describes all of the phases and organizes the sequence of activities and timetable required to develop and commercialize a new variety. It typically covers the full life cycle from new ideas for variety design through to variety registration and discontinuation from seed distribution lists.

Stage plans are used by successful seed companies and public sector R&D institutions to manage their decision making to efficiently and effectively deliver new varieties from their breeding programmes to meet client needs. Organizations may use different nomenclature, headers or activities separated at different points in the process, but the principles are the same, i.e. there is a discrete set of decision points and stages containing defined activities.

The decision points are called stage gates, and these are the points at which formal progression decisions are taken for germplasm to enter the next stage. The separate stages are usually designated by letters, as shown in the list below. This is to avoid any confusion with the crop numbering systems that are used for breeding activities – for some crops, there is a universal numbering system, while for others, there is no consistently recognized nomenclature.

The different stages of a stage plan are as follows:

- Stage A – Variety design;
- Stage B – Trait discovery;
- Stage C – Proof of product concept;
- Stage D – Early development;
- Stage E – Late development;
- Stage F – Pre-commercialization;
- Stage G – Commercialization;
- Stage H – Variety discontinuation; and
- Stage I – Discontinued product.

Figure 5.2 shows a modified version of a stage plan developed by Syngenta Seeds, indicating the decision milestones A to H.

Breeding that is demand led requires the greater involvement not only of farmers but also of all clients and stakeholders in the value chain from the point of inception of the development of new demand-led varieties, through the R&D components of variety creation and breeding, to seed delivery. This greater involvement of stakeholders and broader expert disciplines means that the programme of work has greater complexity, with more data generation and a potential for greater costs, and therefore requires rigorous forward planning and optimal organization for decision making. Therefore, a stage plan is an essential component for delivering a successful and cost-effective demand-led breeding programme.

Taking a demand-led approach as part of the stage plan not only ensures that the variety created delivers the client's needs but also that the development process itself can actively create demand and increase adoption by farmers upon release of the new variety. Creating and delivering demand is achieved by:

- **Consultation:** incorporating clients' ideas and requirements at the design stage.
- **Co-evaluation:** partnering with clients in line evaluation (e.g. using participatory breeding methods with farmers; and the inclusion of tests for suitability by processors for manufacture, chefs for cooking qualities and transporters for storage.

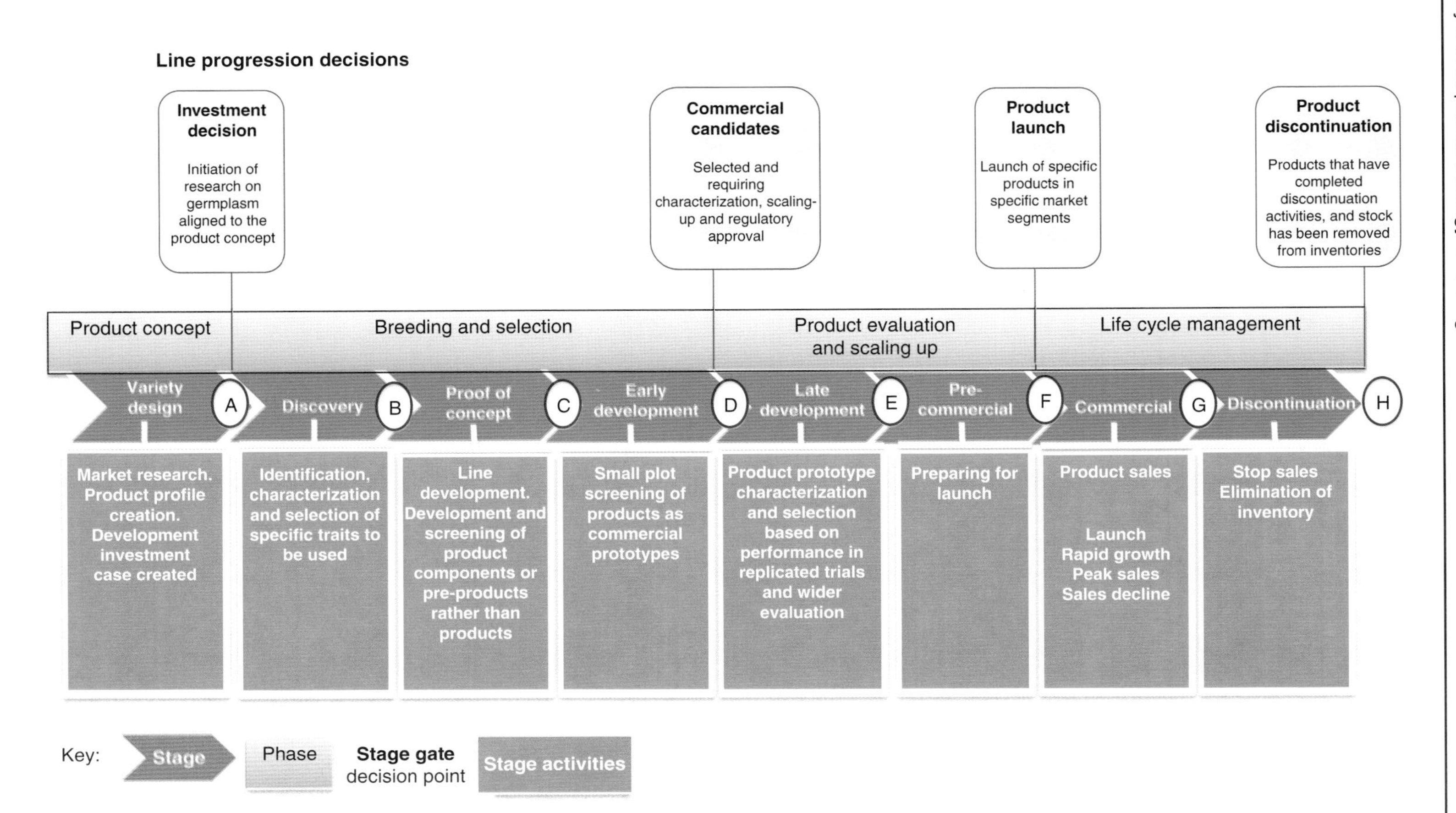


**Fig. 5.2.** A demand-led breeding stage plan. Modified version of a Syngenta Seeds stage plan provided by Syngenta.

- **Joint decision making:** involving clients and stakeholders in the selection and progression decisions of lead lines from one stage to the next.
- **Problem solving:** working with clients and stakeholders to solve problems and make trade-off decisions on the priority of traits in the product profile. When done well, this brings ownership and buy-in by clients.

## Case study: small white pea bean in Zimbabwe

A practical example is provided here that follows the logic of a variety development stage plan for small white pea bean (navy bean) in Zimbabwe (see Fig. 5.3). In this case, the canning foods industry needs a small white pea bean that meets the canned baked bean product characteristics that consumers demand. There is high demand for canned bean products in Zimbabwe, but supply is limited owing to insufficient local production and the lack of white pea beans varieties with suitable canning qualities for farmers to grow. The shortfall is met by imported products from South Africa. The local industry is faced with several challenges in the processing of more locally grown beans, but in addition there is insufficient local supply. As a result, Zimbabwe is importing the white pea bean grain from Malawi and Ethiopia to keep the canning industry going and to meet market demand.

### *Demand*

Zimbabwe's population is approximately 14.6 million (2014) and is increasing at a rate of 3% a year. With the urban population at 33%, and coupled with the rising cost of electricity, more people in urban areas are turning to processed food products to save time and electricity in home food preparation. Canned beans are one of the canned food products that are in high demand. The country therefore needs to develop canning bean varieties that meet the industry requirements, and are also adapted to the bean production environments that are available, are resistant to diseases and drought, and provide farmers with a high grain yield.

### *Stage plan description*

A variety development stage plan for white pea bean in Zimbabwe is described step-by-step in Fig. 5.3. The plan has four phases: (i) product concept; (ii) breeding and selection; (iii) product evaluation and scaling up; and (iv) life cycle management. Within each phase, there steps that are referred to as stages. The different stages are given letter codes A, B, C, D, up to I. At each stage, decisions have to be made before proceeding to the next stage. At the last stage in each phase, critical decisions have to be made and these are called stage gates.

The different phases and stages of a variety development stage plan for this case study are described in more detail in Table 5.1, which describes the phases and stages in sequence and indicates for each their activities, purpose, who is involved in making the decisions, and the next steps.

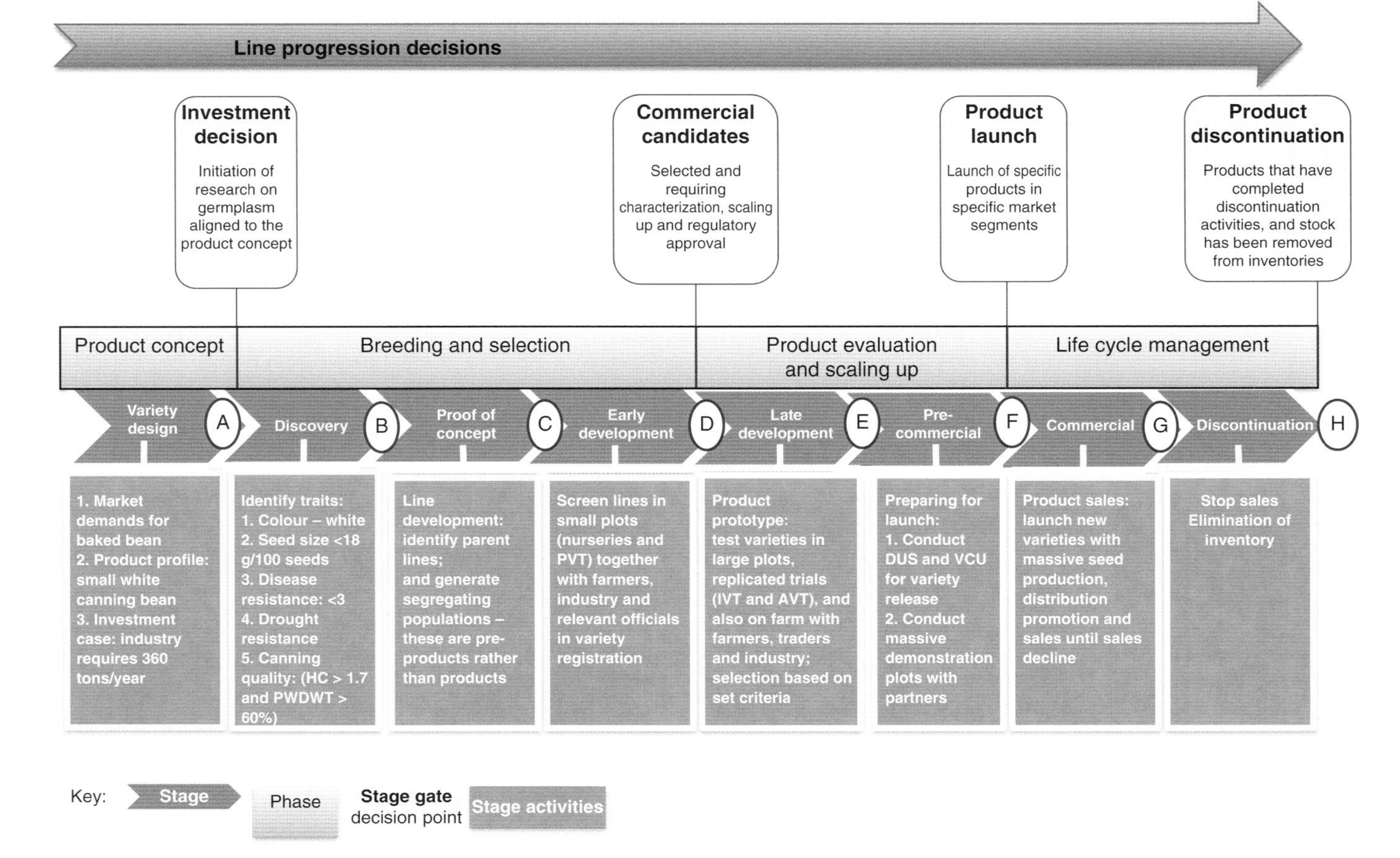


**Fig. 5.3.** A flow chart of a stage plan for developing a new variety of small white pea bean suitable as a canning variety in Zimbabwe. Key: AVT, advanced variety trials; DUS, distinctiveness, uniformity, and stability; HC, hydration coefficient; IVT, initial variety trials; PWDWT, percentage washed drained weight; PVT, preliminary variety trials; VCU, value for cultivation and use. Modified version of a Syngenta Seeds stage plan provided by Syngenta.

**Table 5.1.** Different phases and stages of a variety development stage plan, for the development of small white pea bean varieties in Zimbabwe.

| Stage | Activities | Purpose | Who | Next step |
|---|---|---|---|---|
| **Phase 1: Product concept and Stage Gate 1: Investment decision** | | | | |
| Stage A: Variety design | 1. Develop product profile – describe the canning bean product with clear specifications: Small white pea bean (<18 g/100 seeds, hydration coefficient (HC) >1.7, percentage washed drained weight (PWDWT) > 60%.<br>2. Develop investment case: clearly show need for the product; how much demand is there? (30 tons/month); does cost–benefit analysis favour the investment? | To convince the investors in the public and private sectors, and the clients and value chain stakeholders that the initiative is good value for money | Plant breeder and partners in various fields, including economists and key stakeholders | Proceed to Phase 2: Breeding and selection – initiation of research on germplasm aligned to the product concept |
| **Phase 2: Breeding and selection and Stage gate 2 – Commercial candidate varieties** | | | | |
| Stage B: Discovery | Identification, characterization and selection of specific traits: grain colour (white); grain size (18 g/100 seeds); resistance to diseases (<3.0 score for BCMV[a] and ALS[b]); tolerance to drought; and industrial canning qualities (HC > 1.7 and PWDWT > 60%) | To establish the traits for selection and the selection criteria (cut-off points) | Breeders and the relevant partners, including plant pathologists, physiologists, food scientists, the canning industry and consumers | Proceed to Stage C: Proof of concept |

| | | | | |
|---|---|---|---|---|
| Stage C: Proof of concept | Development and screening of small white pea bean lines for resistance to ALS, CBB[c] and drought, and also for acceptable canning qualities and grain yield under conditions with and without drought stress | To deliver a product that meets the set selection criteria and meets the consumer demand | Breeders and partners (plant pathologists, physiologists, food scientists, the canning industry and consumers) | Proceed to Stage D: Early product development |
| Stage D: Early product development | Small plot screening of products as commercial prototypes | Testing varieties in nurseries and preliminary yield trials (PYT) to select bean lines with resistance to BCMV, ALS and drought, and also preliminary evaluation for yield and canning qualities | Breeder and partners (e.g. variety release committee (VRC), seed certification unit, seed producers and disseminators, grain traders, the canning industry) | Proceed to Stage E, if the product passes the critical evaluation at the end of Phase 2, Stage D (the stage gate on commercial candidate varieties) |
| **Phase 3: Product evaluation and scaling up** | | | | |
| Stage E: Late product development | Product prototype characterization and selection based on performance in replicated trials and wider evaluation | Testing the varieties in advanced yield trials (AYT) or national performance trials (NPT) in preparation for variety release | Same as stage D | Proceed to Stage F |
| Stage F: Pre-commercialization | Preparing for variety launch: conduct large-scale demonstration plots to create awareness of the new variety and release the variety | To create awareness of the upcoming new variety | Same as stage D | Proceed to the stage gate on variety launch (release), after meeting the requirements of Stage F, and then move on to Stage G |

*Continued*

**Table 5.1.** Continued.

| Stage | Activities | Purpose | Who | Next step |
|---|---|---|---|---|
| **Phase 4: Life cycle management** | | | | |
| Stage G: Commercialization | Large-scale seed production and distribution<br>Promotion of the new variety through demonstrations, and mass media (print/electronic) | To publicize the new variety widely and promote sales | Seed producers, seed traders, grain traders and the canning industry | Proceed to Stage I – the last stage towards the stage gate on discontinuation of the variety |
| Stage H: Discontinue the variety | Stop seed production and promotion and deregister the variety from the variety list | To remove the variety from circulation | Breeder(s), seed producers, traders, VRC, and officials in government or in a private seed company | Stage gate on variety discontinuation: replace the old variety with a new, improved variety |

[a]Bean common mosaic virus; [b]angular leaf spot; [c]common bacterial blight.

## The benefits of a stage plan

The benefits to breeders of creating and using a demand-led stage plan include:

- **Product adoption.** Stage plans require regular reviews of the variety design and the trait/germplasm combinations being developed, including engagement with clients and stakeholders as part of these reviews. Thus, there is a higher likelihood of varieties achieving high farmer adoption by meeting clients' demand.
- **Decision making.** A stage plan enables greater focus on the key milestones and generation of the quality data and information that are required for rigorous and informed decisions to be taken in a timely manner during new variety development. It encourages forward planning for convening the governance groups and committees responsible for taking decisions in breeding programmes. It ensures there is management involvement and ownership of the performance of new varieties, and in progression decisions and the associated costs and investment needs.
- **Fully integrated plan.** Stage progression decisions require full transparency of all of the data, information, activities, skills, resources and investment needed to create and deliver a new demand-led variety. This encourages excellent planning to seek efficiencies and a smooth flow of activities from start to completion of the variety development. This is important to avoid delays in or non-delivery of new varieties due to lack of skills, equipment, resources or other programme blockers. As a planning tool, the stage plan also enables iterations with and ownership by all scientific contributors, clients, stakeholders and government authorities.
- **Communications.** The stage plan is a communications tool that provides a visualization framework to show the progress of the many lines under development within a breeding programme or institutional portfolio, as each line is assigned to a specific stage in the plan.

## Creating a stage plan and progression decisions

Every stage plan requires breeders to have a clear idea of how decisions will be taken and what data, information, activities and resources are required, so that key components are in place to deliver the intended new varieties. To create a stage plan a breeder needs to consider and include the following activities: (i) governance and decision making; (ii) engagement with clients and stakeholders; (iii) variety design; (iv) development planning and activity optimization using critical path analysis; (v) breeding, testing and evaluation; (vi) variety registration and scale up; (vii) seed production; and (viii) cost analysis. Overarching all of these activities is monitoring, evaluation and learning (M&E&L), which is required during all stages of the variety development process (see Chapter 6, this volume).

### *Governance and decision making: stage gates and progression decisions*

The lead breeder with their project team, in consultation with their management, should create an appropriate series of stages with decision points, and determine: when these decisions should take place; who is required to make the decisions; what data and inputs are required; and how reviews will be held. The best demand-led breeding programmes create a set of standards to underpin decision making with the specific sets of data, information and performance criteria that are required for a line to formally proceed from one stage to the next. A broad range of information is required, based on: client needs; scientific and technical data; risk analysis of the likelihood of success; and financial aspects, such as resource cost implications. A governance group should be defined that involves key managers, stakeholders, breeding experts and others as appropriate for each stage of progression. This supports buy-in and ownership by management and investors in the programme.

The stage plan reviews will include:

- **Client demand:** reviewing projected demand, client/stakeholder involvement and relationships.
- **Data quality:** assessing whether the data are of the necessary quality and completeness for stage progression or whether further data are required.
- **Germplasm performance:** checking the performance of the lead germplasm or its components versus the product profile and performance required.
- **Risk management:** assessing the uncertainties and likelihood of success and the risk management requirements.
- **Resources and investment:** checking the monies spent, the costs and the availability of skills and resources required for the next stage.
- **Breeding strategy and technical plan:** examining the effectiveness and efficiency of the strategy and plan.

*Engagement with clients and stakeholders*

Decisions are required on which clients and stakeholders should be involved at different points of the stage plan and which ones should be invited to partner in variety design, product profiling, data generation, line progression and stage progression decisions. It is important to bring in constructive ideas from all key clients and stakeholders in the variety development process, including investors, researchers, development partners, seed industry, traders, the processing industry and consumers. These stakeholders will guide which direction the variety development programme should take, clearly define the targets and provide some insights into the economic viability of the programme.

*Variety design*

- Understanding who the new variety will serve and the market segment of farmers and clients for which it is designed.
- Understanding and characterizing the performance and deficiencies of existing varieties.
- Identifying traits to be incorporated in the demand-led variety.
- Identifying possible sources of germplasm containing the demanded traits.

- Actively involving key stakeholders and clients such as researchers, development partners, seed industry, commodity traders, processing industry and consumers in the design process. Clients should also contribute to the decisions on the type of a variety to be developed according to the set criteria that cover both agronomic and market requirements.

*Development planning*

- Capturing investors' interests.
- Establishing the traders' and consumers' preferences.
- Establishing market quality standards and methods of measurement.
- Identifying key players at each stage of the breeding programme and the skills required.
- Estimating the costs of each stage and benefits of the programme.
- Understanding the critical path, timelines and risks.
- Searching for efficiency gains and optimization of the plan and its resources.
- Establishing the monitoring and evaluation approach.

*Breeding and evaluation*

- Creating new germplasm by performing crosses and/or making selections from the resulting best genotypes and testing their performance.
- Identifying key players (personnel, facilities and institutions) that will be required to test and characterize the performance of the new design versus the specification standards required to progress the improved variety through the registration process.
- Defining the outputs and standards of measurements required in assays, testing regimes and field trials.
- Estimating the required resources in personnel/skills, equipment and financial resources.
- Developing a breeding plan of activities from start to delivery of the variety.
- Defining the timelines for each activity.
- Determining the critical path and shortest route to delivery of the variety.
- Defining the monitoring and evaluation process (M&E).

*Variety registration and seed scale up*

In most African countries, the commercial supply of seeds to farmers requires national governments to evaluate and approve the performance and stability of new varieties. Good contact with government officials and key influencers on the variety release committee is essential for all plant breeders so that they understand the registration requirements and the costs involved. It is also important to know the registration timelines for varietal release and to enable the coordination of the scale up of seed production so that seeds can be made available for awareness building and demonstration to farmers as part of the demand-led process.

A key risk is that registration is granted but there is insufficient seed available for farmers to benefit by using the new variety. The stakeholders who need to be involved here are researchers/breeders, development partners, seed traders, the processing industry and consumers. Activities are:

- liaison with registration authorities;
- testing for distinctiveness, uniformity and stability (DUS);
- determining value for cultivation and use (VCU);
- submission to the variety release committee (VRC);
- ensuring the coordination of seed production; and
- conducting M&E.

*Seed distribution and farmer access*

A new variety is only meaningful if the intended users can have access to seeds to enable them to grow the variety. The seed supply systems must be articulated in the development strategy and stage plan so as to ensure farmer access to seed immediately the variety is released. The stakeholders representing the seed supply chain include: researchers, development partners, seed growers, agro-dealers and the private seed industries. It is the seed supply systems that make available and promote the use of the variety and thus determine its possible utilization and adoption, and its ultimate success (or failure).

## Timelines and Critical Paths (Boxes 5.4, 5.5 and 5.6)

Demand-led breeding requires emphasis to be placed on project planning and critical path management. This is to prevent extended timelines and delays that arise from greater consultation requirements and the involvement of clients and stakeholders. The reputations of demand-led breeders will depend not only on the numbers and quality of their publications and the number of new variety registrations, but also on a number of other performance criteria specific to demand-led breeding. These include ensuring that: (i) new varieties are delivering the needs and preferences of farmers and those in the value chain; (ii) new varieties are used by farmers; and (iii) new varieties are developed in a timely and cost-effective manner. Using research approaches that include systematic planning, critical path analysis and risk management will help breeders and their research colleagues to achieve all three of these additional demand-led performance criteria. Learning and deploying these professional skills will help scientists to build their personal reputations for delivery and will contribute to achieving successful careers.

Breeding projects take many years and require a range of different skills from many disciplines, as well as institutes both within and between countries. National and international crop-based networks of breeders are now common and breeding programmes benefit from exchanging germplasm, molecular tools, and ideas and activities all along the development stage plan from the laboratory, to greenhouse trials, to field activities. In addition, more extensive research collaborations occur frequently between universities, national breeding programmes and international research centres. For example, the successful Pan-African Bean Research Alliance (PABRA, 2017), is a network of several national bean research programmes in Africa that is linked with the international

**Box 5.4.** New sugar bean varieties in Zimbabwe: example of a project activity plan.

A bean breeder in Zimbabwe is developing new bean varieties to satisfy the cranberry (sugar bean) market class, which has a wide export market in southern Africa. The market requires a large-seeded grain type (40–45g/100 seeds). In addition, to support the improved nutrition and health for the vulnerable groups (young children and lactating mothers) campaign, the Government of Zimbabwe has a policy that all new bean varieties should meet the bio-fortified bean standards, which means containing at least 90 ppm Fe and 40 ppm Zn (per 100 seeds).

New sugar bean varieties need to meet the following performance standards:

- good environmental adaptation (including tolerance to drought) and high productivity;
- resistance to diseases, e.g. angular leaf spot (ALS), bean common mosaic virus (BCMV); and
- nutritional quality standards of at least 90 ppm Fe and 40 ppm Zn (per 100 seeds).

From more than 200 cranberry lines at the $F_6$ generation harvested in April 2015, the breeder needs to select no more than 30 lines that meet the above selection criteria, and advance the lines for seed increase in the next generation under irrigation during June 2015 planting in the low veld. This crop is targeted for harvest around October 2015 in readiness for planting the preliminary yield trials across four sites under rainfall in January 2016.

To evaluate the Fe and Zn content, the samples have to be sent to Rwanda, where a CGIAR/IFPRI (International Food Policy Research Institute) international programme, *HarvestPlus*, has the XRF equipment to analyse the grain for Fe and Zn content. This has to be done before June, as any delay would mean planting late. Planting after mid-July in the low veld means that the crop's flowering period would coincide with hot night temperatures (above 25°C), which desiccates the pollen, resulting in total crop failure and the loss of germplasm. Given these biological and environmental imperatives, it is important that the variety development team work out all the activities that are needed in a particular year, and then plan the time and order for each one of them, also adjusting when unexpected events occur through risk management. The resulting spreadsheet (or chart) of activities, timelines, resources, partners and the party responsible for ensuring the delivery of each activity on time and on budget constitutes the variety project plan.

bean programme of CIAT (the International Center for Tropical Agriculture) (Buruchara *et al.*, 2011; PABRA, 2017).

These multidisciplinary approaches and public–private partnerships can be magnets for new investments in demand-led variety development. However, such regional and international initiatives are highly dependent on cooperation among research groups located in different countries and, sometimes, in different continents. Successful collaboration requires visionary leadership and close communication among participants, as well as the regular movement of people, breeding materials and equipment around the network, and risk management.

Professional project planning combined with critical path management is one way to create transparency of all the activities needed in a demand-led variety development project, to understand the dependencies, to think about the potential risks and mitigation strategies, and to work out the duration of the project. Such planning is also helpful as a communications tool to enable research collaborators to understand each other's work plans and engender team support.

**Box 5.5.** Variety development project activity plan.

**Year 1a** – Make crosses (Female parent A) × (Male parent B) to get $F_1$ seeds – in the greenhouse – by the breeder.

**Year 1b** – Grow $F_1$ seeds in the greenhouse or in the field under controlled conditions to obtain $F_2$ seeds – by the breeder.

**Year 2a** – Plant bulk $F_2$ seeds in the field under controlled conditions to generate as much seed as possible at the $F_3$ generation – by the breeder.

**Year 2b** – Plant $F_3$ seeds in a targeted environment and select plants with desirable target traits to produce $F_4$ seeds – by the breeder and collaborating partners or stakeholders, depending on the traits of interest.

**Year 3a** – Plant $F_4$ seeds in a targeted environment and select plants with desirable target traits to produce $F_5$ seeds – by the breeder and collaborating partners or stakeholders, depending on the traits of interest.

**Year 3b** – Plant $F_5$ seeds to increase the seed of $F_6$ fixed lines – by the breeder and collaborating partners or stakeholders, depending on the traits of interest.

**Year 4a** – Evaluate $F_6$ lines in a nursery, and select best lines while increasing seed for the next season – by the breeder and collaborating partners or stakeholders, depending on the traits of interest.

**Year 4b and 5a** – Evaluate selected lines in preliminary yield trials (PYT) at three sites and select the best lines – by the breeder and collaborating partners or stakeholders, depending on the traits of interest.

**Year 5b and 6a** – Evaluate selected lines in advanced yield trials (AYT) at four sites, plus on-farm trials at several sites, and select the best lines – by the breeder and collaborating partners or stakeholders, depending on the traits of interest.

**Year 6b and 7a** – Evaluate selected lines in national yield trials (NYT) at five sites and select the best lines – by the breeder and collaborating partners or stakeholders, depending on the traits of interest.

**Year 7b** – Conduct distinctiveness, uniformity and stability (DUS) and value for cultivation and use (VCU) tests – by the designated DUS testing unit.

**Year 8a** – Submit to the variety release committee (VRC) for variety registration (release) – by the breeder and the VRC.

## Demand-led variety development project activity plan

A demand-led variety development project activity plan provides a further level of detail below that of the variety development stage plan and facilitates the timely implementation of the stage plan. A comprehensive and optimized project activity plan is especially important for demand-led breeding. This is because of the increased numbers of people who need to be consulted and the diversity of activities, resources, skills and information required for success. The demand-led variety project plan should provide a description and

**Box 5.6.** Timelines and critical paths: educational objectives.

**Purpose:** to support plant breeders to maximize their chances of successful completion, registration and release of new demand-led varieties on time and on budget, by strengthening their project planning skills, and learning critical path management and risk mitigation approaches.

**Educational objectives:**

- to understand the timelines and costs involved in developing and registering a new demand-led variety;
- to have a clear understanding about critical paths and approaches to risk reduction;
- to develop risk mitigation strategies; and
- to understand the costs and rewards associated with risks.

**Key messages**

- Good project planning ability is a critical skill required by demand-led breeders.
- The duration of each activity requires careful evaluation.
- Data and precedents are used to make decisions.
- Critical path analysis is used to find creative ways to include demand-led testing and consultation requirements without extending timelines and, preferably, by shortening timescales.
- Analysing and preventing risks by anticipating potential problems and introducing critical control points and standard operating procedures (SOPs) is time well spent.

**Key questions**

- How long does it take to design, create and register a new crop variety in your country?
- What are the principles and procedures of critical path analysis and risk mitigation?
- What is the critical path and shortest route to follow in your breeding programme?
- How can you reduce the timelines?
- What are the key risks that can disrupt your programme and lengthen the timelines (e.g. natural weather occurrences, incorrect understanding of the market)?
- What options do you have to reduce such risks (e.g. remnant seeds)?
- How costly and/or rewarding are the risks (e.g. time, risk management)?
- What risks are you prepared to take?
- What is critical control point analysis (CCP) and how can it help to reduce risk?
- What is a SOP and how can it help to reduce risk?

timetable of all activities that are needed during the development of a new variety. The plan is often most effectively communicated as an Excel spreadsheet or a Gantt chart, with all of the activities listed, together with their duration and the person responsible for delivery. The best project plans also map closely on to the development stage plan with milestones, decision points and demand-led performance criteria and standards set to determine whether germplasm lines should be progressed to the next stage. The benefits of rigorous systematic forward planning are demonstrated in the example of sugar bean breeding in Zimbabwe (Box 5.4).

Partnering with the value chain may involve including critical bioassays/genetic evaluation activities to test for the presence of essential consumer traits, agronomic traits, processing traits and/or productivity traits. At each stage and for every activity there are certain skills that are required. If such skills are not available in the home research institution, partnering or outsourcing may be the solution. So it must be clear during the stage plan when certain skills will be required and from where they will be sourced. The timing of activities must be carefully planned, as some must be done sequentially, because they are dependent on each other.

For example, conducting multilocation yield performance trials is done after making selections of the best lines from the segregating populations. Other activities may be able to be done concurrently. For instance, activities such as conducting DUS and VCU tests, could in some countries be done alongside the national performance trials (NPTs), which are conducted to establish the performance of the varieties across different production environments. If it is possible, conducting two evaluation cycles a year shortens the time for variety development. Conducting some activities concurrently can also reduce the timelines and assist rapid delivery of the varieties. Trait evaluation with partners for the demand-led traits can start as early as at $F_3$, depending on the type of trait. However, for some traits (such as canning quality for baked beans), the evaluation must be done much later on fixed lines because the industry requires a larger volume of grain sample to test for this quality.

A variety development project activity plan is illustrated in Box 5.5. This is an example of the activities that are necessary at various stages of a variety development plan, including the trait evaluations to be conducted, when, how and by whom. The activity plan refers to activities to be led by the breeder after the variety product profile has been decided, and the traits of interest have been identified in lead germplasm for use in the breeding programme of a self-pollinated crop. Other examples of project activity plans suitable for crops that are cross pollinated, are hybrids or are vegetatively propagated, are presented by Shimelis and Laing (2012).

## Critical paths

The critical path is the shortest route to complete the full list of the activities from the start of the variety development project to its completion. Understanding and using critical path analysis and risk mitigation planning is especially important in demand-led breeding programmes. These approaches can prevent timelines extending or delays occurring as a result of various factors, including greater consultation requirements and the involvement of clients and stakeholders.

To determine the critical path for a breeding project plan, all of the research, logistics, consultation and information gathering activities need to be listed. The procedure is then as follows:

- Create a start and finish date for each activity.
- Indicate the person responsible for each activity and that they agree with the time required.

- Estimate how long each activity will take.
- Include error bars if unsure of the duration of an activity.
- Identify which activities are dependent on other activities before they can start or for them to be completed. These are called 'dependencies'.
- Map the activities, their time bars and their dependencies into a project plan. This can be done on to a spreadsheet or other chart or, preferably, using a project planning tool.
- Best practice is when each member of the project team provides their list of activities and the lead breeder combines all of these activities into a consolidated project plan, with all of the dependencies highlighted.

This activity time plan can then be used to plot the shortest route (or routes) from the start to completion of the project. This is termed the 'critical path'. There may be more than one path and the path may evolve as activities are completed. At the planning stage, it may emerge that not all activities need to start at the beginning of the project. Deciding whether to do particular research elements, either (i) 'just in time' or (ii) earlier to reduce any risk of unexpected results or overruns, is an important part of the planning process. In the former case, the activity will be on the critical path, with all its inherent risks, while in the latter case, it will not be.

## Dependencies

Dependencies arise when particular activities of the variety development cannot be started or completed without additional information, equipment or results becoming available and/or without another activity first being completed. Often, this introduces a need for researchers to communicate and collaborate with each other, so that the work plan runs smoothly without gaps that lead to delays. Dependencies have risks attached to them. Reviewing the project plan with its critical path, the dependencies and the risks involved is a good way for research collaborators to understand the total research project, solve problems together and realise the benefits of teamwork.

A critical path analysis with risk dependencies is illustrated in Fig. 5.4, as a theoretical example. A practical example of a critical path is the development of new sugar bean varieties in Zimbabwe that are drought tolerant, micronutrient dense and have resistance to two diseases, angular leaf spot (ALS) and bean common mosaic virus (BCMV). The timeline starts with the initiation of the crosses among the parental lines, advancing the segregation populations and making selections among segregating populations and fixed lines, based on set criteria: (i) on grain type – large seeded (40–45 g/100 seeds); (ii) disease resistant (to ALS and BCMV); (iii) drought tolerant; and (iv) meeting the clients' demand for high micronutrient content (Fe > 90 ppm and Zn > 40 ppm). The advanced lines are evaluated in multilocation trials on-station and on-farm, including DUS and VCU, to generate sufficient data for variety release. The variety is then presented to the VRC for approval and registration.

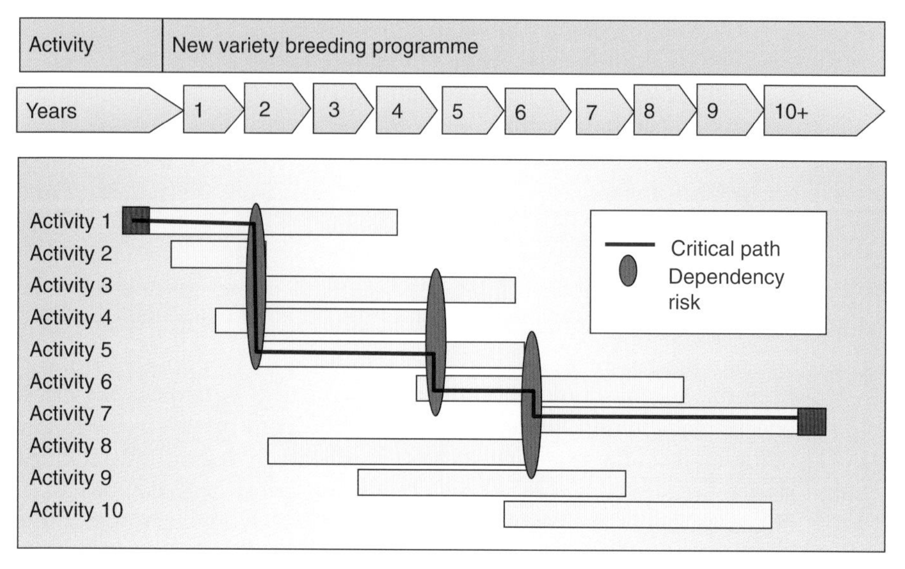


**Fig. 5.4.** Critical path analysis and risk dependencies: a theoretical example. Provided by V.M. Anthony.

This example is summarized in Fig. 5.5, where the activities are listed (numbered) from 1 to 9 and their timelines are drawn across the timescale (10 years). The critical path is highlighted by a red line, which flows from the start to the end points (Fig. 5.5).

The dependent activities (i.e. activities that can only start after other activities are completed) are identified by balloons; they include:

- Activity 2, which starts after the completion of Activity 1;
- Activity 3, which starts after the completion of Activity 2;
- Activity 4, which starts after the completion of Activity 3;
- Activity 5, which starts after the completion of Activity 4;
- Activity 8, which starts after the completion of Activities 5, 6 and 7; and
- Activity 9 after completion of Activity 8.

## Risk Management

There are many problems that can occur during varietal development which can cause delays and, in the worst case scenario, non-delivery. These range from intractable technical and scientific problems, to equipment breakdowns, lack of availability of skilled personnel and funding problems, all of which can affect the critical path to delivery. Also, some scientific activities, by their innovative nature, may provide unexpected results. Where the outcomes of experiments are uncertain and they may require repeating, then it is advisable to schedule them earlier in the plan and not 'just in' time. Risk can also be reduced by using parallel approaches to increase the chances of a successful output, if resources permit.

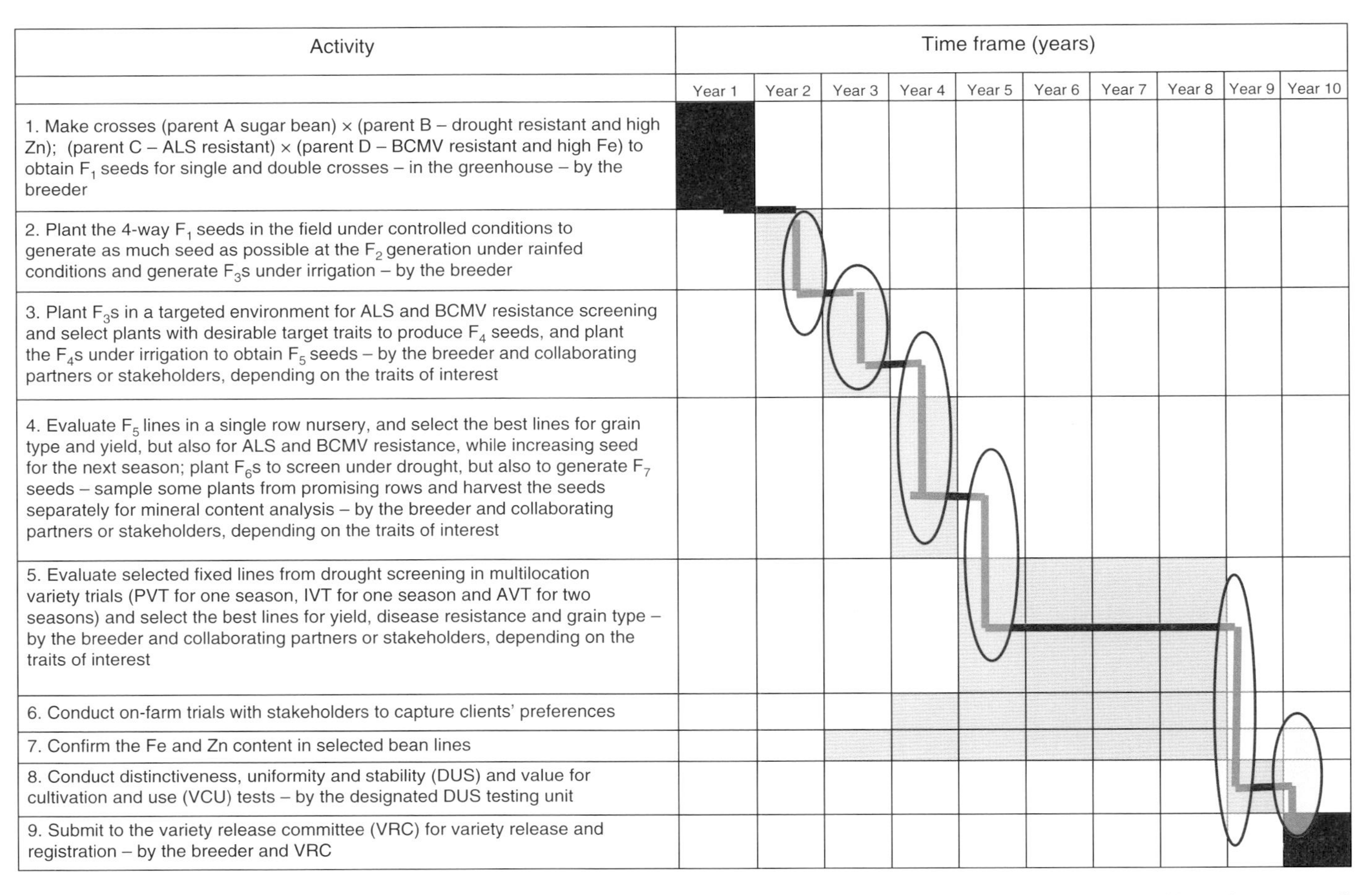


**Fig. 5.5.** Timeline and critical path for developing a new drought-tolerant and micronutrient-dense sugar bean variety with resistance to diseases in Zimbabwe. Key: ALS, angular leaf spot; AVT, advanced variety trials; BCMV, bean common mosaic virus; IVT, initial variety trials; PVT, preliminary variety trials. The blue balloons indicate dependent activities, i.e. those that can only start after other activities are completed.

## Risk identification and mitigation

Best practice in demand-led new variety development programmes uses an approach called 'critical control point analysis' (CCP) as a method to reduce risks. This procedure involves systematically scrutinizing all of the activities required for progression through the stage plan, and places particular emphasis on activities on the critical path.

Each person responsible for delivering part of the project plan identifies activities, handovers to another group, and materials or procedures that could potentially introduce delays or mistakes, or result in the finished variety not meeting the product profile and specification. This is done using lessons learned from previous experiences and anticipating potential problems. Examples of risks include issues during transnational transport of germplasm, occurrence of adventitious presence (i.e. the unintentional and incidental commingling of trace amounts of one type of germplasm with another), computer breakdowns, loss of field trials, etc. Once these risks have been identified, a critical control point should be introduced at which the problem can be prevented or minimized.

SOPs are a good way to prevent errors occurring at a critical control point and to set standards. A SOP is 'an established, written method that describes how routinely to perform a given task successfully'. Case studies on the use of critical control points and SOPs in Africa are given in Johnson *et al.* (2011). While most SOPs are likely to be already in place for routine breeding, some others may need to be created within a demand-led programme; e.g. creating and running bioassay experiments to evaluate a facet of the required variety, or the importation of special equipment or germplasm. Even for those tasks that are done routinely, many may be part of the daily activity but are not well documented. As a consequence, the performance of the same procedure by two different people may differ and that performance may change over time without this being noticed. It is, therefore, important that key procedures relating to demand-led traits are explicitly articulated in written instructions, in an SOP. SOPs are usually written by the person (or persons) that conduct the activity and should be approved and reviewed at regular intervals by the responsible supervisor.

## Risk quantification

Creating a comprehensive list of potential problems, quantifying each risk and ranking (colour coding) its importance can provide rapid assimilation of the collective impact of the individual risks encountered during new variety development. This numerical and visual approach to communicating project risks can help to avoid misunderstandings in interpretation and aids priority setting for the creation of critical control points and risk mitigation actions.

Table 5.2 shows a format that can be used to help think through, assess and communicate risks in demand-led plant breeding projects. Once individual risks in a project have been identified, it is worth remembering that under

**Table 5.2.** Summary assessment of breeding project risks.

| Identified risk | Person responsible | Impact | Likelihood | Risk reduction action |
|---|---|---|---|---|
| Adventitious presence of genetic modification (GM) in crop germplasm | Molecular breeder | Very high | 20% | Passport documentation from germplasm supplier<br>PCR/molecular checks for all GM traits at set points in the stage plan |
| Delayed recruitment of staff | Plant breeder | High | 50% | Organize back-up contractor/ more funds in the budget |
| Lost shipment of seed samples | Seed bank manager | Moderate | 60% | Send seed in two lots, via specialist shipment company |

'probability theory', mathematically you multiply – rather than add – individual chances of success to look at the combined effect of all the risks. So, if there are three major risks with impact, the combined effect is much greater than it may appear. This may affect the next decision and course of action, and it shows the importance of quantifying risks wherever possible.

For example, for the risks in Table 5.2, the combined risk effect is calculated as follows:

**Risk 1.** Very high impact → 20% probability.
**Risk 2.** High impact → 50% probability.
**Risk 3.** Moderate impact → 60% probability.
**Total chance of success.** $0.8 \times 0.5 \times 0.4 = 0.16$ or only 16%!

## Variety Registration (Box 5.7)

It is important that a breeder understands the variety registration requirements, as these will determine whether the variety will be registered or rejected, based on meeting the required standards and procedures. The required processes have a set timeline and that should prepare and guide the breeder to know when the variety is likely to be made available for farmers' use. The requirements to register a new variety and how long it takes to meet the set procedures vary from one country to another. The breeder must understand the regulatory requirements in each country where the variety is expected to be registered.

The regulatory testing requirements and performance criteria for varietal registration of a specific crop are described below, using maize as an example. This is based on a study conducted by Setimela *et al.* (2009), for the International Agricultural Research Centre for Maize and Wheat (CIMMYT). This study showed that, in many maize-growing countries in sub-Saharan Africa (Angola, Benin, Ethiopia, Ghana, Kenya, Malawi, Mali, Mozambique, Nigeria, Tanzania, Uganda, Zambia and Zimbabwe), the registration process requires that after the breeder has conducted trials to identify some promising hybrids or varieties that perform better than the local check variety in one or

**Box 5.7.** Variety registration: educational objectives.

**Purpose:** to develop a clear understanding of the variety registration requirements and the time taken to register a new variety in a specified country or region.

**Educational objectives:**

- to define registration requirements and time to registration for a new improved variety for a crop in a target country or region; and
- to develop a plan to ensure that all registration requirements are met and to know how to engage professionally with variety registration officials.

**Key messages**

- A breeder must understand the time it takes to fully meet the requirements for legally releasing and registering a variety.
- Breeders need to engage early with registration authorities and work with them to achieve registration as early as possible, especially when breeding for new consumer traits that may require new demand-led variety release procedures.
- Breeders should work closely with key government officials, and where possible to sensitize them to the need for new varieties that have traits that respond to market demands – and by so doing, solicit their support for new demand-led varieties so that they advocate their use.
- Where possible, a group of countries can harmonize variety registration regulations to cut costs in registering varieties that can be used by clients across borders. One good example is the harmonized seed regulations for the southern African countries (Zulu and Goldschagg, 2008).

**Key questions**

- What are the regulatory testing requirements and performance criteria for varietal registration for a crop in a particular country and/or region?
- What is the variety release and registration process in a particular country and/or region?
- How long does the registration process take?
- How can one help to speed up the registration process?
- Are there registration costs?
- How will consumer traits that are not recognized by the existing regulatory system be registered?

more traits, that the new varieties must go through tests for DUS and VCU before registration.

The registration process establishes legal ownership of the new variety, according to the seed act of a country. The DUS and the VCU tests can take between 1 and 3 years before sufficient data are available for variety registration, depending on the country requirements. This period is additional to the time it takes the breeder from the stage of generating crosses to the identification of promising varieties out of the AYT or NPT, which can take up to 10 years or more, depending on the number of seasons in a year and the country's regulations.

In each country, a National Variety Release Committee (NVRC) makes a decision to release or to reject a new variety based on the data compiled in the varietal release proposal. In some countries, there is no fee for registering a new variety, but other countries require a fee for the NPT and DUS tests. Each country has its own seed act that determines the variety release regulations, and there is a lot of variability and inconsistency of seed laws across the countries of Africa. This makes it costly for seed companies to release and market new varieties in several different countries, as each variety must be tested every time it is to be marketed in another country, even if it is adapted to a wide range of agroecologies that cut across borders. One way to cut such costs is to harmonize variety registration regulations across countries, where possible.

The regulations on variety testing and release of maize varieties are variable and inconsistent among 13 countries in sub-Saharan Africa. Setimela *et al.* (2009) have suggested that one way to reduce the time it takes to release a new maize variety would be to run the DUS test concurrently with the AYT or NPT, as DUS tests are not affected by the different agroecologies.

Ideally, the release is based on the merit and uniqueness of the variety. As such, the uniqueness of a variety that addresses demand-led traits would be considered sufficient to distinguish a variety from others and that would give it merit for release.

When using demand-led breeding approaches, an issue that requires careful consideration and clarification with national registration authorities is a core requirement for yield improvement. Many VRCs, by following government policies of promoting the agricultural production of staple food crops, will only approve new varieties if they demonstrate yield improvements over existing benchmark varieties. For consumer-based traits, concurrent yield improvements may not be achievable. The need for yield gain must be clarified at the variety design stage and, if possible, buy-in achieved with the pertinent authorities to ensure that registration will be achievable.

### Case study on new maize varieties

The previously mentioned case study by Setimela *et al.* (2009) on drought-tolerant maize for Africa (DTMA) provides insights into the importance of understanding the country requirements for registering new maize varieties in a number of countries in sub-Saharan Africa. It defines the time it takes to release elite maize varieties and summarizes the variety release requirements and procedures in various countries. It also identifies the constraints hampering the release of elite maize varieties to smallholder farmers and proposes strategies to hasten the process to release new varieties.

## Learning Methods (Box 5.8)

Before this chapter concludes, a summary is provided in Box 5.8 of learning methods – together with assignments and assessment methods – for use with the main topics that have been covered in the chapter: New Variety Development Strategy; Development Stage Plan; Critical Path Analysis and Risk Management; and Variety Registration.

**Box 5.8.** Learning methods, assignments and assessment methods.

**New Variety Development Strategy**

*Learning method*

- Presentation on the core elements and creation of a demand-led new variety development strategy, as a planning tool and communications document.
- Group discussions on what a strategy should contain and why each component matters in a demand-led breeding programme.
- Group discussion on understanding the differences between creating a demand-led strategy and a demand-led development stage plan.

*Assignment*

- Each participant should prepare a development strategy and planning document for a variety/group of varieties in their own breeding programme.
- For ongoing breeding programmes, where no strategy exists, it may be appropriate to create a strategy document that includes options for consideration by the programme's management.

*Assessment*

- Assessment of the assignment.
- Exam questions in which the participant demonstrates a clear understanding of all of the elements required to create a demand-led development strategy, why they are important, and the differences between a development strategy, a stage plan and a development activity plan.

**Variety Development Stage Plan**

*Learning method*

- Presentation that shows the key elements of a stage plan and why it is important for a successful breeding programme, together with some practical examples.
- Group discussions, where participants share knowledge and experiences on the importance of a stage plan and risk management therein.

*Assignment*

- Each participant to create a stage plan for their breeding project, which includes: all of the variety development activities required and mapping each activity on to the stage plan; the key decision points/stage gates and the information and results needed to make a line progression from one stage to the next. Also, determine which experts/managers within the breeding project/organization should form the governance group responsible for taking the decision to progress a lead genotype(s) from each stage to the next. This group of responsible individuals may be different for each stage gate progression decision.

*Assessment*

- Assessment of the stage plan assignment.
- Exam questions on the key principles of a stage plan and how to create one.

*Continued*

**Box 5.8.** Continued.

**Critical Path Analysis and Risk Management**

*Learning method*

- Presentation on programme planning and critical path analysis.
- Presentation on risk management and use of standard operating procedures (SOPs).

*Assignment*

- Create an activity plan for your breeding programme/project, with timelines. Describe the critical path and the risk mitigation actions that should be in place during the programme/project.

*Assessment*

- Assessment of the assignment on the project activity plan and critical path analysis.
- Exam questions on the key principles of creating a project plan and critical path and its risk mitigation measures.

**Variety Registration**

*Learning method*

- Presentation on variety registration process and timelines.
- Group discussion on a new variety release with consumer-based traits.
- Visit to a national variety registration unit, with presentations to the group by regulatory officials on their registration requirements.

*Assignment*

- Define the variety release requirements, timelines and costs for a new variety design.
- Visit a member of the variety release committee (VRC) and discuss the breeding goals of your programme and the learning points to take into consideration in new variety design.

*Assessment*

- Assessment of the variety registration assignment.
- Exam questions on the key principles of variety registration and release and why it is important to engage with registration officials on the release of new varieties with consumer-based traits.

## Resource Materials

Slide sets are available for this chapter as part of Appendix 3 of the open-resource e-learning material for the volume. These summarize the chapter contents and provide further information. The e-learning material is available at http://www.cabi.org/openresources/93814 and also on a USB stick that is included with this volume.

## References

Buruchara, R., Chirwa, R., Sperling, L., Mukankusi, C., Rubyogo, J.C., Muthoni, R. and Abang, M.M. (2011) Development and delivery of bean varieties in Africa: the Pan-Africa Bean Research

Alliance (PABRA) model. *African Crop Science Journal* 19, 227–245. Available at: http://www.bioline.org.br/request?cs11022 (accessed 10 May 2017).
Johnson, L., Anthony, V., Alhassan, W.S. and Rudelsheim P. (eds.) (2011) *Agricultural Biotechnology in Africa: Stewardship Case Studies*. Forum for Agricultural Research in Africa (FARA), Accra. Available at: https://issuu.com/fara-africa/docs/sabima_casestudies_ (accessed 10 May 2017).
PABRA (2017) What we do. Pan-Africa Bean Research Alliance, Kampala. Available at: http://www.pabra-africa.org/ (accessed 10 May 2017).
Setimela P.S., Badu-Apraku, B. and Mwangi, W. (2009) *Variety Testing and Release Approaches in DTMA Project Countries in Sub-Saharan Africa*. International Maize and Wheat Improvement Center (CIMMYT), Harare. Available at: http://repository.cimmyt.org/xmlui/bitstream/handle/10883/807/93477.pdf?sequence=1 (accessed 10 May 2017).
Shimelis, H. and Laing, M. (2012) Timelines in conventional crop improvement: pre-breeding and breeding procedure. *Australian Journal of Crop Science* 6, 1542–1549.
Zulu, E.D. and Goldschagg, E. (2008) *Harmonization of Seed Regulations to Promote Seed Trade in the SADC Region: with Focus on Seed Certification; Crop Variety Release and Phytosanitary for Seed Systems*. PowerPoint presentation, FANARPAN [Food, Agriculture and Natural Resources Policy Analysis Network] Workshop, Zambia. Available at: http://fsg.afre.msu.edu/zambia/FANRPAN_workshop/pdf/session_02/E_Zulu_Harminisation.pdf (accessed 10 May 2017).

## Further Resources

Andoseh, S., Bahn, R. and Gu, J. (2014) The case for a real options approach to *ex-ante* cost–benefit analyses of agricultural research projects. In: *Food Policy* 44, 218–226. Available at: http://pdf.usaid.gov/pdf_docs/pnaec758.pdf (accessed 10 May 2017).
AOOC (2017) African Orphan Crops Consortium. University of California, Davis, California/World Agroforestry Centre, Nairobi. Available at: http://africanorphancrops.org/about/ (accessed 10 May 2017).
Artmann, C. (2009) Chapter 2: Literature review. In: *The Value of Information Updating in New Product Development*. Springer, Berlin/Heidelberg, Germany, pp. 9–39. Available at: http://www.springer.com/cda/content/document/cda_downloaddocument/9783540938323-c2.pdf?SGWID=0-0-45-685409-p173876683 (accessed 10 May 2017).
Boettiger, S., Anthony, V., Booker K. and Starbuck C. (2013) *Public–Private Partnerships in Plant Genomics for Global Food Security*. Research paper commissioned from the GATD [Global Access to Technology for Development] Foundation by the International Development Research Centre (IDRC), Ottawa. Available at: https://www.researchgate.net/profile/Sara_Boettiger/publication/242329658_Public-Private_Partnerships_in_Plant_Genomics_for_Global_Food_Security/links/02e7e51cc813353378000000/Public-Private-Partnerships-in-Plant-Genomics-for-Global-Food-Security.pdf?origin=publication_detail (accessed 10 May 2017).
BSPB (2012) Science and Technology: Written evidence submitted by the British Society of Plant Breeders. Lobby paper to British Government on the need for investment in plant breeding. Available at: http://www.publications.parliament.uk/pa/cm201213/cmselect/cmsctech/348/348vw20.htm (accessed 10 May 2017).
Ceccarelli, S. (2015) Efficiency of plant breeding. *Crop Science* 55, 87–97. Available at: https://www.crops.org/publications/cs/pdfs/55/1/87 (accessed 10 May 2017).
GRDC (2011) End Point Royalties (EPR[s]). Fact Sheet, Grains Research and Development Corporation (GRDC), Canberra, Australia. Available at: http://www.seednet.com.au/documents/End%20Point%20Royalties%20Fact%20Sheet.pdf (accessed 22 May 2017).

GRDC (2015) Impact assessment of reports of various crop breeding programmes by the Australian Government Grains Research and Development Corporation (GRDC), Canberra, Australia. Available (some with permission only) at: https://grdc.com.au/about/our-investment-process/impact-assessment (accessed 10 May 2017).

Harries, A. and Cortes, J. (2008) *Procedure Manual for Variety Registration in the National Catalogue for Crop Species and Varieties in West African Countries*. Seed Science Center, Iowa State University, Iowa. Available at: http://www.coraf.org/wasp2013/wp-content/uploads/2013/07/ECOWAS_VAR_REGIST_MANUAL_SEP_081.pdf (accessed 10 May 2017).

Heisey, P.W., Srinivasan, C.S. and Thirtle, C. (2001) *Public Sector Plant Breeding in a Privatizing World*. Agricultural Information Bulletin No. 772, Economics Research Service, United States Department of Agriculture, Washington, DC. Available at: https://www.ers.usda.gov/publications/pub-details/?pubid=42414 (accessed 10 May 2017).

Hickey, L. (2014) The Speed Breeding journey: from garbage bins to Bill Gates. QAAFI Science Seminar, 28 October 2014, Queensland Alliance for Agriculture and Food Information. YouTube video available at: https://www.youtube.com/watch?v=5tsor4PuMmw (accessed 2 May 2017).

Jefferies, S. (2012) Cereal breeding and end point royalties in Australia. Paper presented at the FarmTech 2012 Conference, Edmonton, Canada, as reported by M. McArthur in *The Western Producer* Feb. 9th, 2012. Available at: http://www.producer.com/2012/02/royalty-fee-based-on-production-attracts-breeders-%E2%80%A9/ (accessed 16 May 2017).

Keyser, J.C. (2013) *Opening Up the Markets for Seed Trade in Africa*. Africa Trade Practice Working Paper Series Number 2, World Bank, Washington, DC. Available at: http://documents.worldbank.org/curated/en/916001468008985797/pdf/818340REVISED00o020379857B-00PUBLIC0.pdf (accessed 10 May 2017).

Keyser, J.C., Eilittä, M., Dimithe, G., Ayoola, G. and Sène, S. (2015) *Towards an Integrated Market for Seeds and Fertilizers in West Africa*. World Bank, Washington, DC. Available at: http://www-wds.worldbank.org/external/default/WDSContentServer/WDSP/IB/2015/01/15/000470435_20150115132901/Rendered/PDF/936300REVISED00REVISED0FINAL0TO0DC.pdf (accessed 10 May 2017).

Kuhlmann, K. (2014) *Harmonizing Regional Seed Regulation in Sub-Saharan Africa: A Comparative Assessment*. A paper for seeds2B Africa at the Syngenta Foundation for Sustainable Agriculture, Basel, Switzerland. Available at: https://www.syngentafoundation.org/file/2481/download?token=ascRxaXD (accessed 10 May 2017).

Morris, M.L and Heisey, P.W. (2003) Estimating the benefits of plant breeding research: methodological issues and practical challenges. *Agricultural Economics* 29, 241–252. Available at: http://impact.cgiar.org/pdf/108.pdf (accessed 10 May 2017).

Tripp, R. and Byelee, D. (2000) *Public Plant Breeding in an Era of Privatisation. Natural Resource Perspectives* Number 57, Overseas Development Institute, London. Available at: https://www.odi.org/sites/odi.org.uk/files/odi-assets/publications-opinion-files/2845.pdf (accessed 10 May 2017).

# 6 Monitoring, Evaluation and Learning

Jean Claude Rubyogo[1]* and Ivan Rwomushana[2]

[1]*Seed Systems and Agricultural Technology Transfer, International Center for Tropical Agriculture (CIAT), Arusha, Tanzania;* [2]*International Centre for Insect Physiology and Ecology (ICIPE), Nairobi, Kenya*

## Executive Summary and Key Messages

### Objectives

**1.** To enable breeders to carry out performance benchmarking for their demand-led breeding programmes by devising a realistic performance assessment plan that: (i) is incorporated into strategies and stage plans for new variety design and engages with clients and stakeholders across the value chain; and (ii) develops key performance indicators tailored for new variety development and delivery of the breeding programme's goals and objectives.

**2.** To enable breeders to: (i) appreciate the importance of variety adoption assessment and performance tracking in demand-led breeding; (ii) design pathways for monitoring progress with value chain clients, with defined responsibilities for the various actors; and (iii) explore the use of improved, low-cost methods for variety tracking to monitor the adoption of new varieties.

The aim of the chapter is to enable breeders to design, integrate and implement plans that demonstrate best practices in monitoring, evaluation and learning (M&E&L) in their demand-led breeding programmes, including setting targets based on key performance indicators (KPIs).

The chapter encourages breeders to reflect on what they consider success to look like, both in terms of their demand-led breeding programme and their own professional performance. It focuses on the core principles of demand-led variety design and the best practices for monitoring, evaluation and learning, involving clients in the demand-led process and in the setting of KPIs. It covers

*Corresponding author. E-mail: j.c.rubyogo@cgiar.org

 

the importance, challenges and methods of the post-release monitoring of adoption of new varieties by farmers and value chain clients.

## How does demand-led variety design add value to current breeding practices?

- M&E&L is an important component for success and continuous improvement in demand-led breeding programmes.
- Demand-led M&E&L is designed to support the achievement of set performance targets rather than being driven by breeding activity. It is created at the new variety design concept stage and is an integral part of the development strategy for delivering new and improved crop varieties.
- Demand-led principles and excellent project management form the basis of demand-led M&E&L, and it should be possible to readily incorporate them into existing institutional M&E&L frameworks.
- KPIs of success in demand-led breeding include metrics on: (i) the performance of new varieties; (ii) client satisfaction; (iii) the use of new varieties by farmers and their value chains; and (iv) the numbers of new varieties registered.

*Implications for role of the plant breeder*

- Breeders need to reflect on their own personal goals and support professional performance measurements that include metrics on product use and the performance of new varieties after registration and release.
- They also need to build identity recognition systems into their varieties so that simple, low-cost and, ideally, *in situ* identification can be done to enable variety adoption monitoring.
- Greater engagement and involvement of clients at key decision points in the variety stage plan will help to ensure demand and uptake on the release of new varieties.

## Key messages for plant breeders

*Performance benchmarking*

- **Monitoring and evaluation.** To be successful, demand-led breeding projects require the implementation of best practices in project management, including planning and M&E&L from the new variety design and project initiation stage through to variety release, the widespread use of the variety and its eventual discontinuation.
- **Demand-led strategy and stage plan.** The strategy and stage plan for each new variety design provides the framework, targets, plan and assumptions for all M&E&L activities. Specifically, the stage gates provide the review points for evaluation and learning.
- **Clients and stakeholders.** Engaging key clients in the value chain in the formation of the development strategy and the monitoring and evaluation

process is essential. Specifically, for most new variety development projects, the main clients should be consulted and involved in key decisions at the following stage gates: (i) the decision to invest in the new plant breeding project; (ii) choice of the lead lines to be developed and scaled up; and (iii) new variety release. This will increase ownership of new varieties and ensure longevity of demand.

- **Key performance indicators.** Specific tailored KPIs should be included in the development strategy that supports and encourages the delivery of demand-led plant breeding goals and objectives. Institutional and breeder performance measures may vary and any conflict of interests should be resolved at an early stage with the institutions' research leadership and management.

*Variety adoption and performance tracking*

- **Rationale and benefits.** Breeders need to understand the significance of variety adoption assessment and its links to the breeding programme.
- **Responsibility and funding.** They need to be clear on how adoption will be assessed after varietal release, who is responsible for tracking adoption and whether the breeding programme will require additional funding to enable post-release tracking.
- **Methods and technology.** The use of phenotypic markers or of low-cost, modern molecular technology should be considered to enable more effective tracking of new varieties after release.

## Key messages for research and development (R&D) leaders, government officials and investors

*Performance benchmarking*

- **Development strategy and stage plan.** R&D leaders, managers, officials and investors should encourage and work with breeders to create realistic designs, targets and performance measures, and ensure that these are included in the development strategy, project plan and decision-making process for each new variety.
- **Lessons learned.** R&D leaders, managers and investors should make time for, value highly and share lessons learned within their institutions about how to improve demand-led breeding approaches.
- **Performance frameworks.** R&D leaders and managers should support their staff and find the best ways to integrate the core principles of demand-led performance indicators into their existing performance management frameworks and reporting processes.

*Variety adoption and performance tracking*

- **Variety adoption tracking.** Government officials and investors should support variety adoption tracking with finance, resources, best practices, transparency and encouragement to plant breeders and their clients in the value chain, as a means to improve future performance.

## Introduction

The objectives of this chapter are:

**1.** To enable breeders to carry out performance benchmarking for their demand-led breeding programmes by devising a realistic performance assessment plan that: (i) is incorporated into strategies and stage plans for new variety design and engages with clients and stakeholders across the value chain; and (ii) develops key performance indicators tailored for new variety development and delivery of the breeding programme's goals and objectives.
**2.** To enable breeders to: (i) appreciate the importance of variety adoption assessment and performance tracking in demand-led breeding; (ii) design pathways for monitoring progress with value chain clients, with defined responsibilities for the various actors; and (iii) explore the use of improved, low-cost methods for variety tracking to monitor the adoption of new varieties.

The aim of the chapter is to enable breeders to design, integrate and implement plans that demonstrate best practices in monitoring, evaluation and learning (M&E&L) in their demand-led breeding programmes, including setting targets based on key performance indicators (KPIs). The chapter also aims to act as a resource for education in this field. For this purpose, boxes are included in several sections of the chapter that summarize their educational objectives and present the key messages and questions that are involved. There is also a final box at the end of the chapter that summarizes the overall learning objectives.

The chapter encourages breeders to reflect on what they consider success should look like, in terms of both their demand-led breeding programme and their own professional performance. It focuses on the core principles of demand-led variety design and the best practices in M&E&L, involving clients in the demand-led process and in setting KPIs. It specifically covers the importance of, challenges in and methods for the post-release monitoring of the adoption of new varieties by farmers and value chain clients.

## Performance Benchmarking (Boxes 6.1 and 6.2)

### Definition of terms

There is a series of terms that are commonly used within the plant science R&D and breeding communities in both the public and private sectors, which are used also within M&E&L frameworks. Definitions of these commonly used terms, and examples of their use, are given below. These definitions are advocated for use within demand-led breeding projects.

*Breeding programme*

A breeding programme is 'The combination of all the activities required to create a portfolio of new varieties that serve farmers in different market segments and their value chains'. It comprises the design and delivery of a collection of new varieties for a set period of time. Some of the varieties in the

**Box 6.1.** Example of a breeding goal and its two related breeding objectives for bean breeding.

**Breeding goal:** to create high yielding cultivars with the quality characteristics sought by bean canning factories, thereby improving livelihoods of smallholder farmers and enabling them to enter new markets.

**Breeding objectives:**

- to define and deliver the optimum combination of specific traits and quantitative levels required to meet the quality and taste characteristics required by canners and consumers, so that canning businesses actively seek these varieties to be grown by smallholder farmers; and
- to progress lead lines that perform to the standards required for both farmers and canners, by conducting consultative interviews with farmers and canners before the project starts, and analysing data from factory canning tests and field performance by farmers at appropriate times during the variety development stage plan.

**Box 6.2.** Project outcomes: examples of outcomes of the successful development of new varieties for various clients/market segments.

**Farmers**

- Adoption of the variety by a defined set number of farmers.
- Change(s) in farmers' agronomic practices, such as: earlier planting or harvesting; use of less fertilizer; and/or purchasing more seeds rather than saving their own seed.
- Higher yields that result in farmers choosing to sell their surplus into the local market in addition to growing sufficient food for home consumption.

**Seed retailers**

- Successful performing varieties become in demand and are stocked by more seed retailers.

**Processors**

- Processors recommend that farmers grow specific new varieties in return for guaranteed contract purchasing and prices.

**Plant breeders**

- Feedback information from farmers about the use of the new variety means that breeders change the design of future varieties.
- Breeders undertake increased consultation with farmers and value chain clients to understand their future needs.

**Public and/or private investors (national governments, international development agencies, seed companies)**

- Successful varieties that increase farmer adoption give greater confidence in a plant breeding institute or programme and so more sustainable funding is assured.

portfolio may be varieties for new market segments; they may also be varieties with improved characteristics over those that have been previously released by the breeding programme and are already in use by farmers.

### *Breeding project*

A breeding project is 'The means for creating a new variety to serve the needs of farmers, their markets and value chains'. A project contains the collection of well-sequenced activities (or tasks) in the breeding stage plan. Each activity has a clearly defined objective and duration, and requires the use of inputs such as time, money and human resources to generate an output that satisfies the requirements of the client. Activities may be conducted by breeders and their teams, or by other scientists in their home institutions; they may also be conducted in partnership with farmers, clients, value chain actors and/or other public and private sector organizations. A breeding project has a start and a finish, as shown in the demand-led stage plan diagram illustrated in Fig. 6.1 (see also Chapter 5, this volume).

### *Breeding goal*

A breeding goal is 'The overarching aim that guides decision making within the breeding project'.

### *Breeding objectives*

Each breeding goal has a number of specific 'breeding objectives' that are: 'the measurable steps that can be undertaken to deliver the goal'. The quality and effectiveness of demand-led 'objectives' can be improved by ensuring they have the following SMART characteristics (Wikipedia, 2015), that they are:

- **Specific.** They target specific aspects or combinations of traits that are required for improvement by clients or by the value chain.
- **Measurable.** They can be quantified or at least have an indicator of performance.
- **Achievable.** They can be delivered within the given knowledge, resources and enabling environment.
- **Relevant.** They must relate closely and support the overall goal of the project.
- **Time related.** They are appropriate for the timescales possible and it can be specified when results will be available.

An example of a demand-led breeding goal and its two breeding objectives is given in Box 6.1.

### *Key performance indicator*

A key performance indicator (KPI) is a 'descriptor or metric that can be measured to provide a quantitative assessment of the performance of the core components of the project, such as efficiency and effectiveness of operational delivery, and project outputs, outcomes and impact'.

### *Project outputs*

Project outputs are 'The product(s) that are produced during a demand-led breeding project'. The core output will be the new variety that is designed,

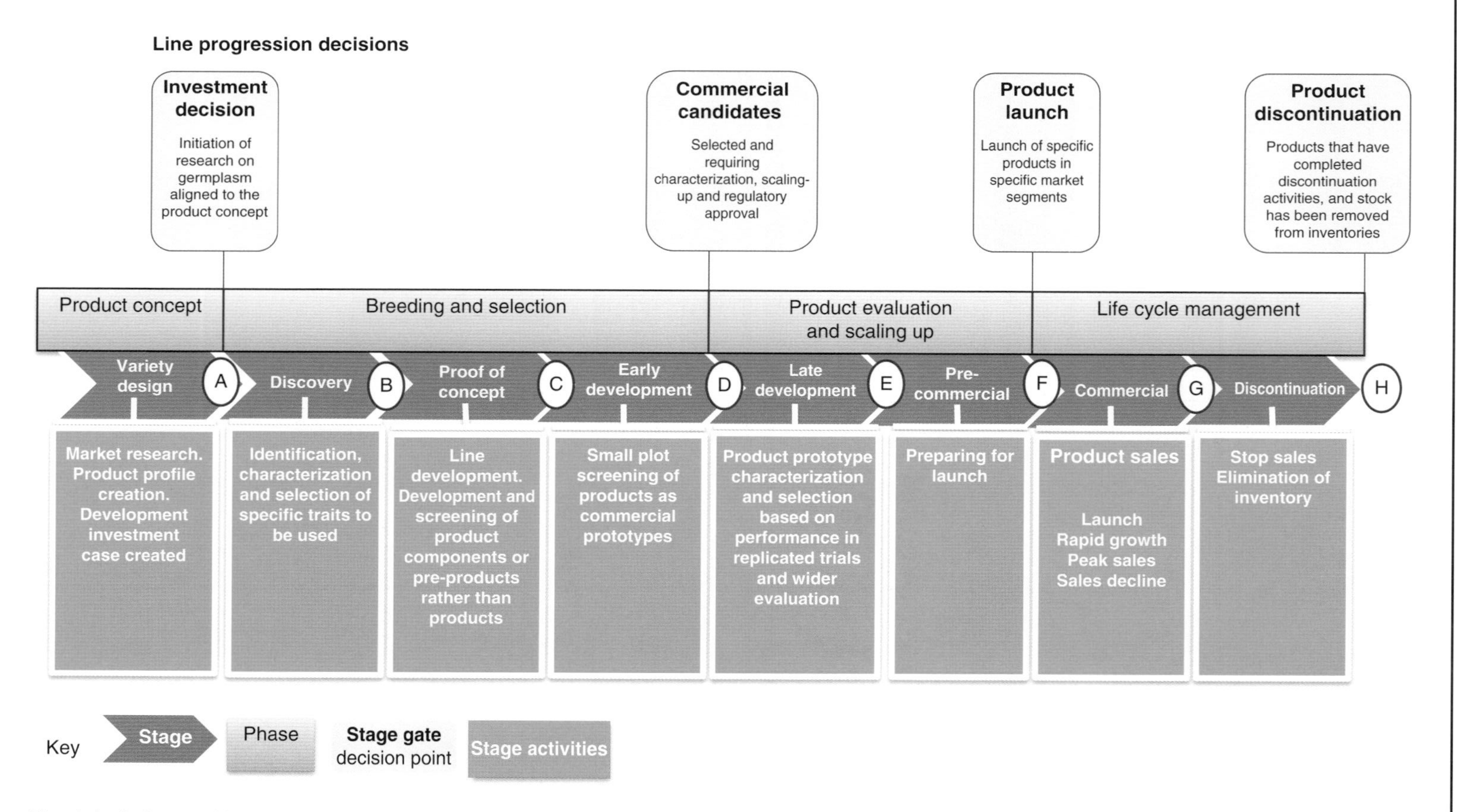


**Fig. 6.1.** A demand-led breeding stage plan. Modified version of a Syngenta Seeds stage plan provided by Syngenta.

developed and released during the project and will be used by clients. Additional outputs include: (i) promotional materials that are used to create client awareness of the benefits and use(s) of the new variety; (ii) diagnostic tools to identify the variety when it is grown in farmers' fields and so support adoption monitoring of that variety; and/or (iii) documented learning in the form of stories of success and failure in the design, development, release and adoption of new varieties, which can be used as case studies for postgraduate education and professional development programmes for plant breeders.

*Project outcome*

The project outcome is 'The changes in behaviour or events that take place as a result of the new variety having been designed, developed, released and made available to clients'. Outcomes can include the effects of project outputs on a range of different clients and stakeholders in the R&D system, market and value chain. Some illustrative outcomes are shown in Box 6.2.

*Project impact*

The project impact is the 'The economic and social benefits that accrue to farmers, their communities and their crop value chains as a result of using the new variety (or varieties) created during a plant breeding project'. The actual impact is measured some years after completion of the breeding project, by means of *ex post* impact assessment studies. Impact assessment helps to track the long-term impact and wide-ranging changes that a breeding project brings about to deliver benefits to its clients. Impact assessment focuses on changes beyond those visible or achieved during the lifetime of most breeding projects.

Other, *ex ante* impact assessment studies may be conducted earlier, either before or during a breeding project, to predict the likely impact of new varieties on the target clients. *Ex ante* studies can be helpful in making the business case for investment in a new breeding project. *Ex ante* impact studies may be used also to assess the effectiveness of R&D interventions, and investors may use them to assess the likely benefits accruing from their investments in breeding projects.

Demand-led breeding is target and investment/return driven and so assumptions on predetermined benefits are set at the beginning of the project and evaluated throughout the development plan, and after varietal release and adoption of the new variety, so as to progressively monitor achievements and likely benefits.

*Target*

A target is 'A quantitative expression of the desired result of the breeding project. It is usually a numeric result(s) and can describe an output or an outcome, and be a key performance indicator'.

*Variety branding*

Variety branding is 'A name, term, design, symbol or other feature that distinguishes one institution's/company's/individual breeder's variety from those of others'. It occurs during the pre-commercialization stage and before full commercialization. The aim of variety branding is to distinguish the variety from others by highlighting the unique characteristics/features of that particular variety. A description of these unique characteristics usually appears on the seed packages used to market the variety and/or may be printed on flyers/posters and other marketing and communication tools.

An understanding of variety branding is becoming more important for breeders as a means of ensuring variety protection. The complexity in nomenclature, the ease of propagation and the numerous levels of distribution and promotion of new crop varieties necessitate protection. The breeding and commercialization of new crop varieties is an important economic activity for research institute(s), breeder(s) and seed company(ies), and for the food and agricultural sector of society. So protection for new varieties through branding ensures that companies, public institutions and independent breeders receive returns on their investments of time and money, and so can justify their involvement in the research and development of new plant varieties. Branding also enables breeders to be recognized for their skills and makes breeding a valued profession. An example of a variety brand for the bean variety JESCA is shown in Fig. 6.2. In this

IMPROVED BEAN VARIETY JESCA

| Characteristics | Benefits |
|---|---|
| • Thrives well in low and medium altitudes | • High yielding |
| • Matures early (70–82 days) | • Highly marketable at prime prices |
| • Yield levels (8–10 bags/acre) | • Disease resistant – increases farmers' incomes |
| • Cooking time (40–48 minutes) | • Ensures food security (leaves and grains) |
| | • Improves soil fertility |
| | • Crop residues can be used as livestock feed |
| | • Can be intercropped with other crops, such as maize |

FOR HIGHER YIELDS OBSERVE GOOD AGRICULTURAL PRACTICES

- Prepare the field well
- Do not burn the crop residues
- Plant early at a spacing of 50 cm × 20 cm
- Use good quality seed
- Apply fertilizer
- Weed at the right time

**Fig. 6.2.** An example of variety branding for a new bean variety, JESCA, by the Agricultural Seed Agency (ASA) of Tanzania.

example, the Agricultural Seed Agency (ASA), a Tanzanian producer and supplier of high-quality agricultural seeds that are available to farmers at affordable prices, has the exclusive right to market the new variety.

## Principles and Best Practices in Demand-Led Breeding

Superlative plant breeding programmes comprise a portfolio of breeding projects, each with clear goals, specific objectives, outputs, outcomes and impacts. They drive the creation of new varieties and form the framework for monitoring, evaluation, impact assessment and learning. There are three principles that are required for success in demand-led breeding: a target-driven approach, a demand-led variety development strategy and performance indicators.

### Target-driven approach

Demand-led breeding is target driven and this approach is embedded deeply throughout the breeding project/programme and the accompanying monitoring, evaluation and learning processes. Great emphasis is placed on quantitative goal and target setting in order to enable improved varieties to reach the clients they are designed for and to fulfil client expectations.

*Best practices in demand-led breeding*
This target-driven approach is exemplified by the following best practices in demand-led breeding:

- **Variety design.** A detailed list of traits with quantified levels of performance is defined and compared with the traits of the existing varieties for line progression to take place.
- **Client quantification.** Numbers of farmers, their location, market segments and targeted clients in value chains are quantified at the outset of the project.
- **Variety adoption.** Adoption levels by farmers are set and monitored for success. Variety registration is an important part of the process to enable farmers to access a new variety.
- **Development stage plan.** A time plan of activities to generate the data required to make line progression decisions is created before the start of the breeding project. The timing of inputs by clients and managers towards the making of these decisions is determined as part of the stage plan and decision points (see Fig. 6.1).

### Demand-led variety development strategy

A demand-led variety development strategy is designed for each new variety and includes all of the components of 'what', 'why', 'who', 'when' and 'how'.

The strategy contains a stage plan for line progression decisions, together with a set of development activities and an investment plan for delivery. Monitoring and evaluation (M&E) is an integral part of the project delivery plan, and a set of KPIs is included in the strategy to provide targets for evaluation. This strategy and its components are used as the baseline for all M&E work.

The quality of the strategy is determined by: (i) visioning and foresight about market opportunities and client demand; (ii) engagement with clients to seek their feedback at key decision points in the stage plan, from new variety design through to post-release variety impact evaluation; (iii) ensuring policy coherence and alignment with the country's national priorities and the enabling policy environment; (iv) realism on the cost, benefits and appropriateness of investment in the breeding programme; and (v) well-designed, technically feasible operational plans for the creation and delivery of each new variety that serves clients in particular market segments and agro-ecological zones (see Chapters 1–5, this volume).

### Performance indicators

The level of engagement and emphasis placed on the views of clients on the performance and use of varieties is much higher in demand-led than in conventional breeding programmes. The success of a new variety and its KPIs is determined by the opinions, demand and use of the new variety by farmers and clients within the crop value chains.

Successful demand-led programmes satisfy end-user demand and are highly dependent on the assumptions formed during investigative research and collaboration with clients and stakeholders along the value chain. These assumptions form the strategic pillars for M&E&L during the development, release and adoption of new varieties by farmers. The required engagement with clients and value chain stakeholders, and the creation and delivery of a variety development strategy and stage plan, will require discussion and approval by senior management.

These core principles of demand-led breeding may not require a specialized M&E framework to be developed for demand-led breeding projects if the principles can be integrated into existing institutional performance management frameworks and M&E processes.

## Monitoring, Evaluation and Learning for Demand-led Breeding

M&E is a process to improve project performance and achieve results, and it is applicable to any breeding programme. The main reasons for including M&E and systematic learning and communication as a component of a demand-led breeding programme are:

- **Delivery:** to support the development and delivery of new varieties that meet clients' needs, fit their design specification (variety profile), and are delivered on time and on budget.

- **Quality:** to provide data on project progress and effectiveness; to improve project management and decision making; to provide data to plan future resource needs; and to provide data useful for policy making and advocacy.
- **Accountability:** to ensure accountability to clients, public and private investors, partners, value chain actors and other stakeholders.
- **Learning:** to provide opportunities to learn from experience in current breeding projects and the overall breeding programme; and to provide evidence about what works and what does not, so as to inform future projects and the scaling up of new varieties.

## Monitoring

Monitoring is a continuous observation and checking procedure that involves gathering information on the progress of ongoing breeding activities. It includes comparing the progress of activities against the milestones and timelines for decisions in the breeding development stage plan and comparing the estimated and actual costs of the project against its associated budget.

The purposes of monitoring are: (i) to support reaching the milestones and targets set for the project in a timely manner; (ii) to determine whether corrective action is required to solve emerging problems or any delays; and (iii) to identify potential improvements that need to be made to the plan.

The breeder, project leader or implementer of each activity checks whether it is going as planned and executes adjustments in a methodical manner. For each new variety design, the project leader/breeder, in conjunction with their management, needs to decide on *who*, *how* and *when* the monitoring of activities will occur, and the frequency of reporting on the scheduled review of all parts of the activity plan. This monitoring plan and reporting schedule should be included as part of the development strategy document that is approved by the breeder/project leader.

## Evaluation

Evaluation is a systematic and objective examination of the performance of a breeding project and the delivery of its goals, objectives and targets. The best evaluations consider project performance in terms of relevance, effectiveness and efficiency, and whether the expected outputs, outcomes and impact have been or will be achieved on time and to budget.

Evaluation is an intermittent procedure that is scheduled at various times during the project to support delivery of the objectives and the targets, for example: at key milestones such as stage gates when the project moves from one phase to another; in response to a particularly critical issue; and, usually, at the end of the project to review actual delivery and the learning that can be applied to future projects.

In most demand-led breeding projects, there are four key times when critical evaluation needs to be undertaken to support investment and progression decisions. These key decision points are as shown in the stage plan (Fig. 6.1), when decisions are to be made on:

- investment – the decision to start a new breeding project and invest in creating a specific new variety design;
- commercial candidates – decisions on which lead lines are to be developed and scaled up;
- new variety release; and
- post-launch adoption and impact assessment.

At each evaluation stage, engagement and consultation with clients, value chain actors and stakeholders is required: (i) to ensure that assumptions on demand and variety performance are reviewed and validated; and also (ii) to check that there is continued support from R&D management and government officials for continued investment. In some organizations, objective evaluation, especially after the completion of a project, is only considered valid if independent, external M&E professionals conduct the review.

For each demand-led breeding project, the project leader/breeder, together with their teams, clients and the management, should define the key indicators to be measured and used used in the evaluation to assess progress in achieving the main outputs, outcomes and impact of the project at the levels of the individual, household and community. Appropriate tools and procedures for monitoring these indicators need to be devised, each with a clear timeline.

Ultimately, a breeding project should be evaluated in terms of:

- **Meeting trait performance targets.** How close is the performance of the new variety to the original new variety design/specification targets set at the product profile/concept stage? The original design would have been determined using visioning/forecasting methods, market research and inputs from clients and the value chain. More specifically, are the genetic improvements required for each of the traits being delivered?
- **Satisfying clients' needs.** Does the new variety satisfy the client's needs and demand? Is it preferred to older varieties? Has it been adopted by the target numbers of farmers for whom it was designed?
- **Impact.** Does the new variety create the economic, social and environmental impact at the individual, household and community level that was defined in the benefits case that was used to justify the investment in the breeding project? (Note that impact can only be assessed several years after variety release, in order to allow time for its adoption).

## Key questions for M&E

Some key questions to consider as part of the M&E system for a demand-led breeding project are:

- Do you have a clear, feasible and documented new variety development strategy?
- Do you know who and how many clients your variety is being designed for and who is likely to use it once it is registered?

- Have farmers and other clients along the value chain been consulted adequately for their inputs during the new product design stage and at key points in the development process?
- Has demand been adequately quantified?
- How clear and deliverable are the project goal and objectives?
- Is the new variety product profile fully defined, with each trait having a target performance benchmark?
- How well defined is the investment benefits case and how robust are the assumptions?
- Has a stage plan been created containing clear decision points for line progressions, and does it include inputs from clients and stakeholders at each key decision point?
- How comprehensive and integrated is the development activities plan that supports the stage plan?
- How feasible is the product profile and its delivery on the set timelines?
- Did the variety deliver the target performance for the clients it was designed for? Were the promised genetic gains achieved?
- What are the internal and external factors that may influence project implementation and delivery (e.g. technical, managerial, organizational, institutional, socio-economic and/or political risks)?
- Have all of the key risks been defined and mitigation actions included in the development delivery plan?
- Was a realistic budget set that was closely aligned with the breeding activities plan?
- Was the project delivered on time and on budget?
- Was variety registration achieved?
- Was seed system production, and also the route for farmers to access the new variety so that demand could be satisfied adequately, addressed in the development strategy?
- How effective were the partnerships selected for the delivery of the new variety?
- How well and accurately could breeders or other stakeholders identify the new variety during the post-release monitoring processes?
- Did farmers adopt the new variety at the levels and pace expected?
- How could greater multiplier effects and a larger scale of adoption be reached?
- Did the farmers and the value chains gain the social and economic benefits expected (as they were set in the business case to justify the investment in the breeding project?
- What learning experiences and recommendations came from the project that will inform future breeding projects?

## Project performance management

Performance measurement is one of the cornerstones for success in any breeding project. It ensures that measures are aligned to the development strategy and that the monitoring system is working effectively. There is a range of different methods for performance management that are suitable to use in demand-led breeding projects. These include:

- **Quantitative key performance measures.** These are numerical measures, gathered by using, for example, surveys with set questionnaires, and/or predefined data collection formats, which are useful for recording and comparing predetermined variables.
- **Qualitative descriptions.** The change is shown through description; for example, by rapid assessment techniques, focus group discussions and/or semi-structured interviews.
- **Participatory methods.** These include visual methods such as photos, diagrams, maps, timelines, stories, drama, song, mapping, ranking and scoring. These approaches allow the differences in people's interests, needs and priorities to be recognized by insiders and outsiders, and form the basis for negotiation between stakeholders. They also allow people to benefit from analysing and asserting their own interests, and to make a meaningful contribution to the way the project is designed, managed and implemented. Participatory methods are a way of ensuring relevance of the project and promoting ownership by local communities, which is essential for sustainability and success in the long term.
- **Most significant change (MSC).** This is one of the most popular approaches that have been developed for M&E without indicators. It is a form of participatory M&E, based on telling stories about events that people think were important and why they think they were significant. With this approach, there is no need to explain what an indicator is or to learn special professional skills, so everyone can participate and the technique can be used in different cultural contexts. MSC is particularly suited to monitoring the project where the focus is on learning as well as on delivery.
- **Outcome mapping.** This technique focuses on changes in the behaviour and attitudes of people, groups and organizations that are directly affected by the development or release of the new variety. For example, outcome mapping focuses on the effect of the project on the roles and participation of partners from the value chain for this and future projects.

## Key performance indicators

The focus in designing a suitable M&E system for monitoring performance in demand-led breeding projects/programmes is to have rigorous, target-based KPIs. The M&E system should use methods that maximize quality, reliable and accurate feedback from clients, value chain actors and stakeholders. It should also include conventional scientific measures of breeding efficiency and effectiveness. In particular, KPIs need to be created that measure client benefits, satisfaction and varietal adoption over time. Breeders need to define and agree the appropriate performance indicators with their teams, with the institutional management and with their clients, value chains and other stakeholders.

A successful breeding project is one that produces outputs, outcomes and impact. These three project components can each be affected by the efficiency and effectiveness of the activities and project management used for operational delivery. In demand-led breeding projects, KPIs are recommended to be created and monitored for each of these three project components. Some KPIs are likely to be the same as those used in conventional breeding programmes, while others are more specific to a demand-led approach. Indicators need to be selected so that they provide a precise description of an objective to be measured, and are realistic, verifiable, and relevant to the clients and value chains that are creating demand.

Examples of indicators to measure performance of the various project components within demand-led breeding projects are outlined below.

*Output indicators*

- Development strategy – indicators of quality, deliverability and management support.
- Variety design – index or measurement scale of farmer interest, support and demand.
- Variety performance versus target profile – percentage technical performance, client satisfaction rating.
- Variety identification – percentage reliability/accuracy of the detection method.
- New variety promotion materials – measurement of quality and ease of understanding by extension professionals.
- Number of new varieties registered and released.

*Outcome indicators*

- Numbers of farmers growing the new variety.
- Area of the new variety grown (as total area and as percentage of total crop area in country).
- Number of farmers multiplying seed of the new variety for sale.
- Number of processors advocating use and seeking farmer production of the new variety.
- Number of seed retailers/organizations distributing the variety.
- Client enthusiasm to partner with plant breeders in new variety design and development.

*Impact indicators*

- Increased yield per farm area.
- Gross margin and incremental profitability of a tonne/unit area of harvested crop for farmers, traders and processors.

- Economic benefit to cost ratio for all players in a value chain.
- Social benefits for women and farming communities.

*Operational delivery*
- Actual cost expenditures versus approved budget.
- Timelines delivered on time or within x% of expected deadlines.

### Learning and communications

Learning is an important aspect of a breeding project that is often overlooked. Learning from experience and gaining knowledge of how to more efficiently and effectively implement a demand-led breeding project has significant value. This value can be described both in terms of enabling the continuous professional development of plant breeders and improving the delivery of new, demand-led varieties through investments in future breeding projects.

Success stories and case studies are important communication tools to share learning and impacts among breeders, clients, stakeholders and investors. Such case studies provide a medium for understanding, remembering and acting on knowledge gained and lessons learned. They also provide a lasting record of the accomplishments of the breeding team in developing new varieties that meet market demand and have impacts that last well beyond the completion of a specific breeding project.

Lessons can also be learned from less successful breeding projects, but this is more challenging. It requires visionary R&D leadership and supportive management to ensure that the evaluation of less successful breeding projects leads to positive outcomes – outcomes that contribute to the professional development of plant breeders and also guide future new variety designs and investments in new breeding projects towards greater success.

Box 6.3 outlines the educational objectives of the topics that have been covered in the last three main sections: **Performance Benchmarking, Principles and Best Practices in Demand-led Breeding, and Monitoring, Evaluation and Learning for Demand-led Breeding**.

## Variety Adoption and Performance Tracking (Box 6.4)

### Varietal adoption studies

Numerous studies show that the economic benefits accruing from investments in crop genetic improvement are large and higher than their opportunity costs (Echeverría, 1990; Alston *et al.*, 2000; Evenson, 2001). Several impact assessment studies have also shown that the returns from plant breeding research are large, positive and widely distributed, and that the gains in benefits that emanate

**Box 6.3.** Performance benchmarking, principles, best practices, monitoring, evaluation and learning for demand-led breeding: educational objectives.

**Purpose:** to enable breeders to devise a fit-for-purpose performance assessment for their demand-led breeding project.

**Educational objectives**

**Performance measures:** to understand the process and key performance indicators (KPIs) used in your home institution to evaluate the success of the breeding projects and your performance as a plant breeder; to be able to critique these measures for their suitability for use in demand-led breeding projects and develop additional KPIs as required.

**Demand-led performance management:** to be able to design and lead performance management of the breeding project, using best practices in monitoring, evaluation and learning (M&E&L) in the context of demand-led breeding; and, specifically, to be able to evaluate the success of your breeding project(s) in terms of relevance, efficiency, effectiveness and impact.

**Value chain partners:** to understand the importance of the participation of clients and value chain actors in M&E&L systems and to know how and when to seek their views and involve them.

**Definitions and linkages:** to understand the terminology used in project management and M&E&L, and the progression from inputs to activities to outputs to outcomes to impacts.

**Monitoring progress:** to use the new variety development strategy and stage plan as the baseline to monitor progress towards the demand-led project breeding goals, objectives and performance targets.

**Learning:** to create a mechanism for learning and taking appropriate collective action on improvements.

**Success stories**: to be able to write success stories about demand-led breeding projects.

**Communications:** to be able to communicate monitoring and evaluation (M&E) results, success stories and learning from demand-led breeding projects and programmes, and to foster greater engagement with clients, value chain actors and other stakeholders on the use of new varieties.

**Key messages**

- Applying best practices in project management and M&E&L will increase the performance and likelihood of success in breeding programmes.
- Demand-led breeding is target driven and the development strategy for each new variety design and its associated stage plan should be used as the baseline reference for all M&E&L assessments.
- The breeding goals, objectives, target clients and market segments that the new variety is being bred for, and how the new variety will be developed, must be clear before a successful M&E&L plan can be initiated.
- Specific KPIs and methods for seeking input and feedback on performance from clients and their value chain actors should be included in the development strategy to support and encourage the delivery of plant breeding goals and objectives.
- It is important to discuss and agree with research leaders and senior management the proposed new variety designs, the associated development strategy and the KPIs that will be used to support the delivery and assessment of the breeding project. Any conflicts or modifications required with an institution's existing performance management system need to be resolved with senior management before the breeding project is initiated.

*Continued*

**Box 6.3.** Continued.

- Key clients should be consulted on at least four key decision points (or stage gates) in the stage plan: (i) the decision to invest in the demand-led plant breeding project; (ii) the lead lines to be developed and scaled up; (iii) new variety release; and (iv) variety adoption and impact assessment studies.
- Sharing learning experiences during and after completion of the development of new varieties is a valuable and constructive part of improving breeding projects and programmes. Special focus is required in demand-led breeding to strengthen professional development, teamwork, partnerships and collaborations with clients and value chains.

**Key questions**

- What terminology do you use to describe your breeding project, its outputs and the benefits and consequences for the farmers who use your varieties?
- What do the following terms mean: breeding project, breeding programme, output, outcome and impact?
- Does your institution have an established M&E&L system that you use to monitor progress in the conduct of your breeding programme?
- What are the strengths and weaknesses of the current M&E&L system?
- How will your institution measure your individual performance as a breeder?
- How will you measure your own performance as a breeder?
- How will you build/expand on an excellent reputation as a professional breeder?
- What does success look like to you?
- Who are you breeding your new variety for?
- What are the overall goals and objectives of your breeding project?
- What are the breeding targets for your new variety?
- How will you measure progress towards the targets and what assessment tools will you use?
- What is a key performance indicator (KPI)?
- Which KPIs are most appropriate to measuring the success of your new variety?
- Are the KPIs for demand-led breeding programmes similar to or different from those now used in your home institution's M&E&L system; if they are different, how?
- What is the difference between monitoring and evaluation, and why are they both important?
- Why is setting a baseline a fundamental requirement for successful M&E?
- How can your new variety development strategy and stage plan help you to monitor and evaluate progress and success in your demand-led breeding project?
- How will you track progress in your breeding project?
- Why is it important and what are the benefits of involving clients, value chain actors and stakeholders in your M&E&L system?
- How and when will you seek client and value chain inputs and feedback into your M&E&L system? What will be the challenges to gain their continuing engagement?
- What feedback mechanisms will you put in place to enhance effectiveness in your breeding programme and increase variety adoption?
- To what extent has a results and performance measurement-based framework been employed in your planning process?
- How will you share your results with others and increase the learning process from your experiences in breeding and variety use?

**Box 6.4.** Variety adoption and performance tracking: educational objectives.

**Purpose:** to enable breeders to understand the importance of variety tracking and the use of best practices in gathering reliable data on farmer adoption and variety performance after release.

**Educational objectives:**

- to evaluate and select the most appropriate variety tracking methods, in conjunction with clients and value chains; and to be able to use these methods in setting measurable criteria for success in variety adoption, as part of a monitoring and evaluation (M&E) and impact assessment system for demand-led breeding; and
- to be able to appreciate and develop learning and communication tools to discuss and share the results of new variety development, adoption and performance with clients and stakeholders.

**Key messages**

- Demand-led breeding projects are driven by understanding and satisfying client demand. Being able to measure new variety adoption, client satisfaction and the resulting economic and social benefits is an essential part of the development strategy and stage plan for new variety development.
- Breeders need to discuss with and gain commitment from their management about how varietal adoption will be assessed after release, who will be responsible for this, and that sufficient resources will be available for this purpose.
- Impact assessments will be required after completion of the breeding project to evaluate the effects of the project on the lives of farmers and other clients, using measures from each key performance indicator (KPI) target.
- The KPIs are likely to include assessment of the following:
  - What is the rate of adoption of the new variety and the associated output products?
  - How many farmers (farm households) will/have benefited directly (or indirectly) from the new variety and your interventions?
  - How much land (hectares) has been planted with new varieties?
  - What economic and social benefits have clients received?
  - What changes in behaviour have taken place and what benefits have resulted?
- Breeders need to: be aware of all the variety identity and tracking technologies available (e.g. phenotypic as well as low-cost molecular approaches); evaluate each option in terms of accuracy, technical feasibility and cost; and select and incorporate the most appropriate one(s), into the new variety development strategy and plan.

**Key questions**

- Why will you assess the variety use and adoption?
- How do you conduct the variety adoption study/assessment?
- How will you know if your variety has been adopted?
- How will you or others be able to identify your variety in the field?
- How do you know if you have achieved the standard required?
- Do you need to develop an assessment method for use by government officials or clients so that they can monitor the performance of consumer-demanded traits in new varieties (e.g. nutrition levels)?
- How can you conduct participatory variety adoption (PVA)?
- How can you assess farmers' variety satisfaction, market variety share and other clients' satisfaction?
- How does variety adoption assessment guide the future breeding programme?

from the adoption of improved crop varieties among communities stretch beyond the favoured regions to marginal areas, and are spread reasonably well between producers and consumers (Heisey and Morris, 2002).

Depending on the underlying breeding objectives, improved varieties may have an array of benefits, including higher yields, improved output quality, lower production costs, simplified crop management techniques or shorter cropping cycles. However, these assertions can only hold if farmers take up and utilize the varieties generated through plant breeding research. Further, the success of the breeding programme can only be verified if the adoption pathways and performances of the products are tracked and documented over time.

In estimating the benefits of a breeding programme, several approaches may be used, including: (i) measuring the area under improved varieties; (ii) estimating farm-level yield gains; (iii) accounting for changes in crop management practices; (iv) accounting for non-yield benefits, such as improved grain quality; (v) accounting for policy distortions; (vi) considering price effects in output markets; (vii) attempting to translate farm-level yields into aggregate supply response; and (viii) considering the *without-project* scenario.

Documenting the outcomes and impacts associated with investments in plant breeding has been a long-standing challenge in Africa. Despite the many efforts and large investments that have been made in developing and releasing improved crop varieties, measuring the levels of uptake of these varieties, understanding the true drivers of adoption, assessing the scale of adoption, and estimating the various impacts derived from variety adoption, have been incomplete and difficult tasks. These challenges in demonstrating the outcomes and impacts of new plant varieties in Africa have hindered research organizations and investors from fully demonstrating the relevance and magnitude of the benefits of their investments in plant breeding.

One of the most comprehensive recent studies on new variety adoption in Africa has been the Diffusion and Impact of Improved Varieties in Africa (DIIVA) study conducted by Walker *et al.* (2014) for the CGIAR Impact Assessment Panel. This study examined the adoption by farmers of some 1150 new varieties of 20 crops in 30 countries of sub-Saharan Africa over 15 years (Walker and Alwang, 2015) (see also Chapter 1, Table 1.1, this volume). The rates of adoption ranged from almost 90% for soybean to less than 7% for banana. The average rate of adoption of modern varieties across all crops in all countries surveyed was 35% in sub-Saharan Africa. This contrasts with approximately 65% variety adoption in Asia and 80% in Latin America.

## Constraints to tracking variety adoption

A major constraint to determining variety uptake has been the difficulties in identifying crop varieties once they are in farmers' fields. Most farmers in Africa access new seed of preferred varieties through predominantly informal seed systems, including local seed markets. In these informal channels, it is difficult to keep track of the origin of the planting material and new names for varieties start to multiply, making it difficult to identify the actual variety in use.

For instance, growing beans as variety mixtures is a common practice in many African countries, and often leads to the loss of identity of different varieties. With the wide diversity in market class (seed colour, size and shape) of beans, and uncertainty in the identification of the actual varieties being used by farmers, adoption and impact studies may provide incorrect estimates of the actual use of improved varieties, and hence a misleading analysis of the determinants of adoption and its associated impacts. The confusion over variety identification also poses a problem in utilizing the newly released varieties and indigenous landraces as a genetic resource in breeding programmes, especially when reporting and comparing experimental results.

The use of DNA fingerprinting technology to complement household surveys in varietal adoption studies can reduce the number of varieties not identified or misidentified in farmers' fields. These new molecular data can provide a clearer picture of planted materials, and when layered with data from household surveys, can provide valuable information about the drivers for adoption, and link farmer-level data to the socio-economic landscape. Where formal seed distribution systems are operational, retailers who are supplying clean seeds to farmers, together with seed producers who are growing seeds of new varieties for sale to retailers, are also important sources for gathering information on adoption and client needs.

## Benefits of tracking variety adoption

An informed basis on which to guide public and private breeders on the characteristics and performance of varieties that farmers, their communities and value chains want in practice, and to drive breeding programmes, depends on the following factors:

- better understanding of variety use and the factors affecting adoption, especially in informal seed systems where accurate seed production data are limited;
- estimation of the extent of the national and regional spread of improved varieties to inform current strategies on regional trade within specific market corridors;
- enhancement of cross-border seed and grain trade, especially for multiple released varieties;
- determination of the cost-effectiveness and efficiency of current genotyping platforms for fingerprinting analysis;
- clarification of the extent of genetic diversity in crops that are in use, and the related risks to biotic and abiotic stresses;
- identification and conservation of new and unique germplasm to protect agri-biodiversity; and
- support for new variety dissemination projects that include the accurate identification of varieties or certified seed, so as to ensure seed purity.

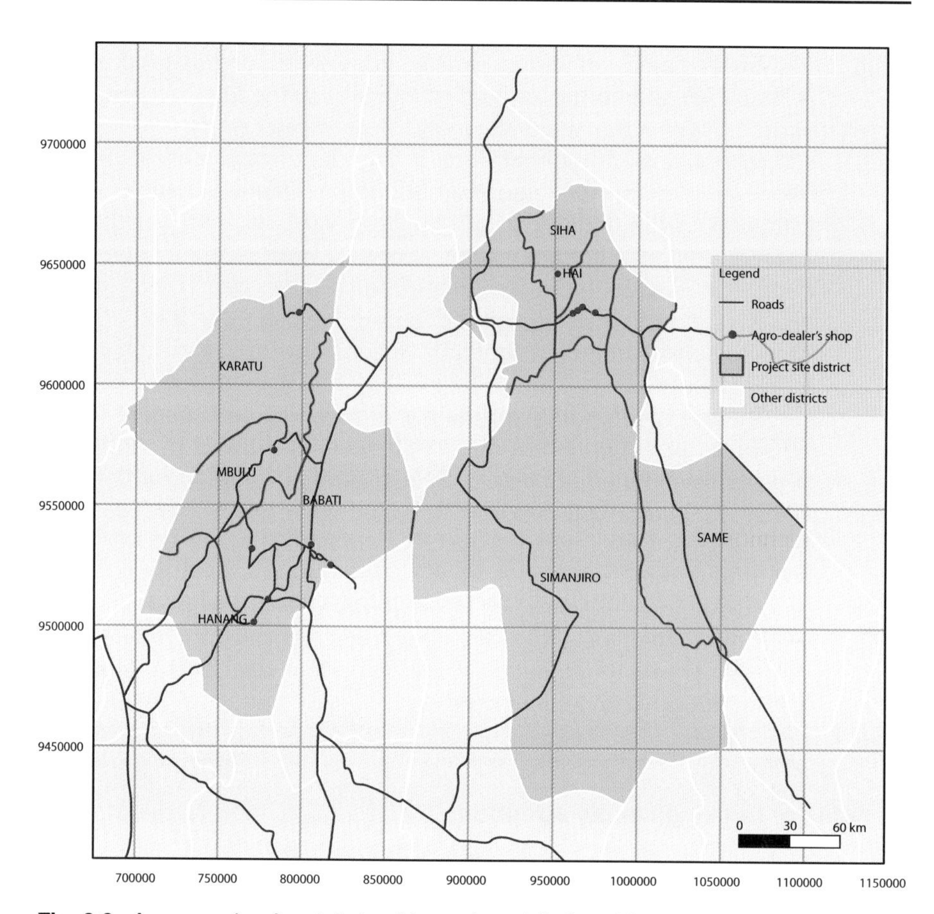


**Fig. 6.3.** An example of variety tracking using global positioning system (GPS) coordinates in six districts of Tanzania. The map shows the location of agro-dealers' shops in Babati, Hanang, Mbulu, Karatu, Siha and Hai.

### Variety tracking using GPS coordinates

In demand-led breeding, institutions, companies and individual breeders should, as a necessity, monitor the movement of their varieties and determine whether their new varieties are in the main growing locations, and are being widely adopted and used by farmers. This information should also include the locations of sales agencies such as agro-dealers, as well as their geographical distribution. An example of variety tracking using GPS coordinates is shown in Fig. 6.3.

## Case Study of the Bean Sub-Sector in Ethiopia, 2004–2014

This case study documents the changes driven by new varieties in the Ethiopian bean sub-sector before and after 2004. It demonstrates a range of important factors contributing to new variety adoption and the importance of breeders understanding their clients and value chains.

## Bean production and marketing constraints in Ethiopia before 2004

Constraints to bean yield and productivity in Ethiopia before 2004 included low levels of input use (fertilizer, pesticide, improved seed) that were attributable to low farmers' incomes, low levels of irrigation, soil degradation, inadequate agricultural research and extension, and constraints in market development. For several decades, bean research in the Ethiopian Institute of Agricultural Research (EIAR), including its Regional Agricultural Research Institutes (RARIs) and research undertakings in universities, was coordinated by the EIAR bean research programme based in Melkassa. The aim of the programme was to respond to the existing bean production constraints by generating production technologies targeting increased bean productivity for enhanced household food security in Ethiopia.

However, there was limited interaction among the bean value chain actors, despite the increasing interest in beans as a marketable commodity by traders and exporters. The link between the producers and the export markets was weak, owing to the large number of ineffective intermediaries operating in the value chain. The small-scale intermediaries operated in limited geographic areas, fragmented between the producer and consumer markets, and creating a lack of transparency in bean marketing. The large (bean) importing companies could seldom manage thousands of smallholder bean farmers who also operated under non-standardized production systems, and lacked market forecast information, adequate financial systems and basic logistics.

In terms of bean seed supply, there was limited access for farmers to fresh seed from the either the research organizations or seed suppliers. The main source of certified bean seed of improved varieties was a government parastatal organization, the Ethiopian Seed Enterprise (ESE). For instance, in 2004, the ESE supplied about 40 tons of bean seed – less than 1% of the national seed requirement in that year. The ESE was supplying mainly three varieties of white pea bean out of 21 released up to 2003 by the EIAR, the RARIs and Hallemaya University (Assefa *et al.*, 2006). The ESE seed was supplied primarily to government-supported seed intervention projects. The *woreda* (district) bureaux of agriculture were then responsible for the distribution of seed to the farmers. There was also limited decentralized seed production by farm research groups supported by research centres around their experimental stations. The rate and extent of adoption of the released varieties remained quite low (Assefa *et al.*, 2005).

## Bean seed access, production and marketing in Ethiopia, 2004–2014

A more robust seed systems approach was introduced in Ethiopia after 2004 and yielded remarkable results. The new Ethiopian bean seed system is illustrated in Fig. 6.4.

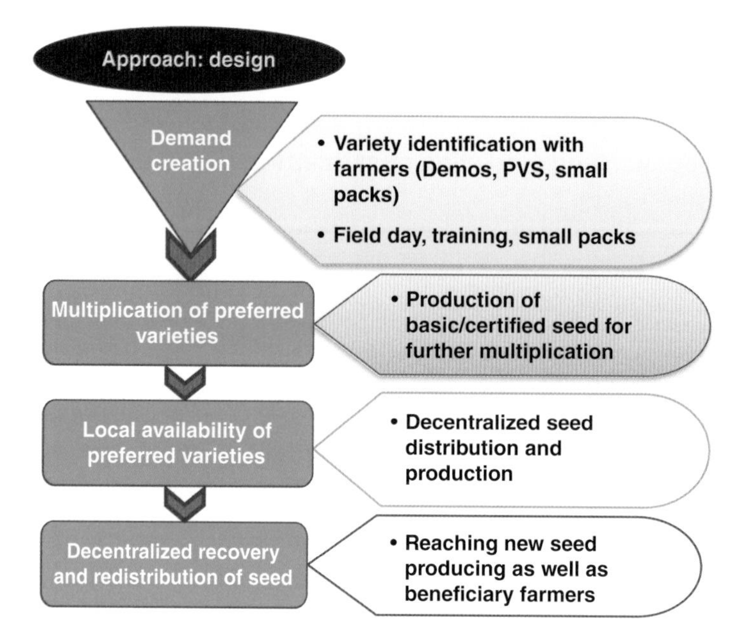


**Fig. 6.4.** Bean seed access, production and marketing in Ethiopia, 2004–2014. Key: Demos, demonstrations; PVS participatory varietal selection; small packs, small and affordable bean seed packs.

Between 2004 and 2014, a sustained, diversified and decentralized bean seed supply dramatically increased access by famers to seed of improved bean varieties. The annual review meeting of the Ethiopian bean multi-stakeholders indicated impressive impacts on bean production resulting from the novel targeting of demand-led varieties, and engagement with decentralized seed producers, who were supported by partner organizations (including non-governmental organizations – NGOs, farmer cooperatives and district agricultural offices).

The adoption and use of small, affordable bean seed packs expanded the reach of new varieties, especially in remote areas where new variety penetration was previously low. The approach proved efficient in enhancing access to high-quality bean seed by resource-poor farmers, particularly women (ICRISAT, 2011). Access to seeds of market-demanded varieties increased from less than 20% to about 65% across major bean growing areas between 2004 and 2010 (Katungi *et al.*, 2009; Katungi *et al.*, 2011). The number of partners also increased from 15 to 53 between 2004 and 2014. In the same period, the amount of seed produced increased 500-fold and the number of varieties promoted by the partners increased 200-fold.

The wider use of improved bean varieties was tied to bean pricing and increased market demand. The Ministry of Agriculture facilitated the exposure of farmers to market-demanded varieties and complementary technologies. Between 2004 and 2014, there was a 33% increase in area under beans – from 245,507 to 326,500 ha, while production increased by 115% from 211,347 to

455,115 tons. Bean production increased as result of both significant increased area under beans and increased grain productivity, with productivity increasing from 0.861 tons/ha to 1.49 tons/ha between 2004 and 2014.

Decentralized bean seed production improved the livelihoods of farmers through increased and diversified incomes, which helped them to meet their financial needs, including those for school fees and household items (Gebeyehu and Dadi, 2010; IPMS, 2010). Some farmers were able to reinvest the proceedings from bean sales to expand the area under bean production and/or to invest in livestock production (Teshale *et al.*, 2007). More exporters have joined the bean trade and are creating employment, especially for women. For example, Agricultural Cooperative Supplies (ACOS Ethiopia) employs more than 1200 women and runs a nursery school where children learn while their mothers are working.

The significant changes in the bean sector have been attributed to: (i) increased incentives (better prices) along the supply (value) chain; (ii) increased adoption and use of improved bean varieties and complementary crop management technologies (for example, by increasing the frequency of weeding from one to three times a season); (iii) increased access to improved technology packages, including improved postharvest management handling techniques; (iv) increased awareness of farmers and other supply chain actors about the linkages between seed quality and grain quality; and (v) the prioritization of beans by government policy makers and crop research management and their decision to support the development and strengthening of the bean value chain (Rubyogo *et al.*, 2010).

Overall, the transformation of the bean sub-sector in Ethiopia has been significant over the past decade for six major reasons:

- enhanced access to improved varieties and complementary crop management technologies;
- effective partnerships;
- supportive government policies;
- an improved bean marketing process and value chain, rewarding higher prices to farmers;
- value chain connectivity; and
- a multi-stakeholder approach.

To achieve the improved value chain connectivity, continuous efforts were made through national and regional platforms to engage actors in creating bean value chain fora at the local level and engaging through these in discussions at the national level. Stronger linkages between exporters and smallholders create more efficient and effective bean value chains where demand signals are clearly communicated to the producers, and inputs are available to ensure production of the requisite export volumes. More involvement of reliable private sector partners in the production of basic seeds will further improve seed availability and accessibility to sustain the momentum as the demand increases.

The multi-stakeholder approach adopted in the transformation of the bean sector in Ethiopia needs to be both continued and expanded in order to engage an even wider range of stakeholders to transform the bean value chain in a more holistic manner. Collective action among stakeholders is required to fully transform the whole bean value chain. This includes local and national government bodies, development partners and investors, private companies, local and global agricultural companies, and financial institutions. Partnership coordination units have been established to engage a broad network of stakeholders in Ethiopia.

Political will is necessary in strengthening national agricultural plans, facilitating the formulation of policies aimed at improving the enabling environment for local bean traders and exporters, and increasing public investment in agriculture-related infrastructure to reduce transaction costs for small-scale bean entrepreneurs.

The private sector should also increase agriculture sector-related investment with an emphasis on developing sustainable, innovative and smallholder-inclusive business models so as to enhance the efficiency and effectiveness of the bean value chain. Finally, local companies should engage actively in partnerships to increase access to capital, which will boost the capacity of smallholder bean entrepreneurs and incentivize their efforts towards bean commercialization.

## Learning Methods (Box 6.5)

Before this chapter concludes, a summary is provided in Box 6.5 of learning methods – together with assignments and assessment methods – for use with the two main topics that have been covered in the chapter: Performance Benchmarking; and Variety Adoption and Performance Tracking.

**Box 6.5.** Learning methods, assignments and assessment methods.

**Performance Benchmarking**

*Learning method*

- PowerPoint slide presentation, including definition of terms, principles and best practices in demand-led breeding.
- Group discussion on the performance management system used by institutions for monitoring and evaluation (M&E) in demand-led breeding programmes, and the challenges in its implementation.

*Assignment*

- To review a prototype new variety development strategy (as created in Chapter 5, this volume), and to construct or improve the M&E plan created for your demand-led breeding project. Highlight the principles used and indicate the main steps you will undertake to track and evaluate progress and learning. This includes defining: who is responsible, who is consulted and who undertakes the M&E and learning activities; when and how will these be done during the development stage plan and after project completion; how information from M&E and learning will be communicated and to whom, and what tools will be used?

*Continued*

**Box 6.5.** Continued.

- Select and define a set of key performance indicators (KPIs) and measures that are most appropriate to assess the success of your performance and that of your breeding project and explain why. The KPIs need to relate closely to your project's goals, objectives, outputs, outcomes and impact.

*Assessment*

- Assignments as above.
- Exam to test understanding of: the definitions, principles and best practices in monitoring, evaluation and learning (M&E&L) relating to demand-led breeding; who, how and when to involve clients in demand-led breeding projects; and the differences between demand-led and conventional breeding projects.

**Variety Adoption and Performance Tracking**

*Learning method*

- Group discussion on how to identify your variety in the field.
- PowerPoint slides outlining issues on variety adoption and performance tracking.
- Case study on the transformation of the bean sub-sector in Ethiopia 2004–2014.

*Assignment*

- Design the role of the breeder in a variety adoption study.

*Assessment*

- Assignment.
- Exam to test understanding of the importance and challenges that breeders face when conducting variety adoption studies.

## Resource Materials

Slide sets are available for this chapter as part of Appendix 3 of the open-resource e-learning material for the volume. These summarize the chapter contents and provide further information. The e-learning material is available at http://www.cabi.org/openresources/93814 and also on a USB stick that is included with this volume.

## References

Alston, J.M., Chan-Kang, C., Marra, M.C., Pardey, P.G. and Wyatt, T.J. (2000) *A Meta Analysis of Rates of Return to Agricultural R&D. Ex Pede Herculem?* Research Report 113, International Food Policy Research Institute, Washington, DC. Available at: http://ebrary.ifpri.org/cdm/ref/collection/p15738coll2/id/125334 (accessed 15 May 2017).

Assefa, T., Abebe, G., Fininsa, C., Tesso, B. and Al-Tawaha, A.-R.M. (2005) Participatory bean breeding with women and small holder farmers in eastern Ethiopia. *World Journal of Agricultural Sciences* 1, 28–35.

Assefa T., Rubyogo, J.C., Sperling, L., Amsalu, B., Abate, T., Deressa, A., Reda, F., Kirkby, R. and Buruchara, R. (2006) Creating partnerships for enhanced impact; bean variety delivery in Ethiopia. *Journal of Crop Science Society of Ethiopia* 12, 1–19.

Echeverría, R. (1990) *Public and Private Investments in Maize Research in Mexico and Guatemala*. CIMMYT Economics Working Paper 90/03, International Maize and Wheat Improvement Center, Texcoco, Mexico. Available at: http://libcatalog.cimmyt.org/download/cim/21017.pdf (accessed 15 May 2017).
Evenson, R. (2001) Economic impacts of agricultural research and extension. In: Gardner, B.L. and Rausser, G.C. (eds) *Agricultural Economics, Volume 1A, Agricultural Production*. Elsevier/North Holland, Amsterdam, pp. 573–628.
Gebeyehu, S. and Dadi, L. (2010) *Improved Common Bean Production and Post-harvest Handling Practices in Ethiopia*. Catholic Relief Service, Addis Ababa.
Heisey, P.W. and Morris, M.L. (2002) Practical challenges to estimating the benefits of agricultural R&D: the case of plant breeding research. Prepared for presentation at the 2002 Annual Meeting of the American Agricultural Economics Association (AAEA) Long Beach, California, July 28–31, 2002. Available at: https://core.ac.uk/download/pdf/6407477.pdf (accessed 11 May 2017).
ICRISAT (2011) Leading a legacy of legumes. In: *ICRISAT 2011 Annual Report*. International Crops Research Institute for the Semi-Arid Tropics, Hyderabad, India, pp. 8–9. Available at: http://oar.icrisat.org/6040/1/J120_2012AnnualReportsmall.pdf (accessed 11 May 2017).
IPMS (2010) *Improving Productivity and Market Success (IPMS) of Ethiopian Farmers Project: Monitoring and Evaluation Report of Year 5 (2009/2010)*. IPMS Ethiopia, International Livestock Research Institute (ILRI), Addis Ababa.
Katungi, E., Farrow, A., Chianu, J., Sperling, L. and Beebe, S. (2009) *Common Bean in Eastern and Southern Africa: A Situation and Outlook Analysis of Targeting Breeding and Delivery Efforts to Improve the Livelihoods of the Poor in Drought Prone Areas – Under the Auspices of Objective 1 of the Tropical Legumes II, Funded by the Bill & Melinda Gates Foundation, Through ICRISAT*. International Center for Tropical Agriculture (CIAT), Kampala. Available at: http://tropicallegumes.icrisat.org/wp-content/uploads/2016/02/rso-common-bean-esa.pdf (accessed 11 May 2017).
Katungi, E., Horna, D., Gebeyehu, S. and Sperling, L. (2011) Market access, intensification and productivity of common bean in Ethiopia: a microeconomic analysis. *African Journal of Agricultural Research* 6, 476–487. Available at: http://www.academicjournals.org/article/article1380904171_Katungi%20et%20al.pdf (accessed 11 May 2017).
Rubyogo, J.C., Sperling, L., Muthoni, R. and Buruchara, R. (2010) Bean seed delivery for small farmers in sub-Saharan Africa: the power of partnerships. *Society and Natural Resources* 23, 285–302.
Teshale, S., Dumetre, A., Darde, M.L., Merga, B. and Dorchies, P. (2007) Serological survey of caprine toxoplasmosis in Ethiopia: prevalence and risk factors. *Parasitology* 14, 155–159.
Walker, T. and Alwang, J. (eds) (2015) *Crop Improvement, Adoption and Impact of Improved Varieties in Food Crops in Sub-Saharan Africa*. CGIAR Independent Science and Partnership Council (ISPC) Secretariat, Food and Agriculture Organization of the United Nations (FAO), Rome, and CAB International, Wallingford, UK.
Walker, T., Alene, A., Ndjeunga, J., Labarta, R., Yigezu, Y., Diagne, A., Andrade, R., Andriatsitohaina, R.M., de Groote, H., Mausch, K. *et al.* (2014) *Measuring the Effectiveness of Crop Improvement Research in Sub-Saharan Africa from the Perspectives of Varietal Output, Adoption, and Change: 20 Crops, 30 Countries, and 1150 Cultivars in Farmers' Fields*. Report of the Standing Panel on Impact Assessment (SPIA), CGIAR Independent Science and Partnership Council (ISPC) Secretariat, Food and Agriculture Organization of the United Nations (FAO), Rome. Available at: http://impact.cgiar.org/sites/default/files//pdf/ISPC_DIIVA_synthesis_report_FINAL.pdf (accessed 15 May 2017).
Wikipedia (2015) SMART criteria. Available at: https://en.wikipedia.org/wiki/SMART_criteria (accessed 15 May 2017).

# 7 The Business Case for Investment in New Variety Development

ROWLAND CHIRWA*

*CIAT (International Center for Tropical Agriculture), Chitedze Agricultural Research Station, Lilongwe, Malawi*

## Executive Summary and Key Messages

### Objective

- To strengthen plant breeders' ability to create compelling business cases for investments in demand-led plant breeding.

The chapter describes the elements necessary for plant breeders to be able to create a compelling plant breeding investment proposal to R&D management, government officials and financial investors. This includes identifying the benefits and intended beneficiaries of a proposed new breeding project/programme, understanding the principles of a return on investment; and clarifying whether the investment in demand-led breeding can be justified in terms of the economic, social and environmental benefits versus the costs.

### How does demand-led variety development add value to current practices?

- **Benefits and investment cases.** Greater emphasis is placed on analysing and creating compelling business cases. This is done by identifying and communicating the full breadth of quantitative and qualitative economic, social and environmental benefits that will become available for clients and stakeholders by investing in the proposed demand-led plant breeding programme/project.

*E-mail address: r.chirwa@cgiar.org

 

*Implications for the role of the plant breeder*

- **Market and business knowledge.** To be effective at consultation with the value chain, demand-led breeders require greater knowledge about crop uses, markets and the economics of breeding. The demand-led approach integrates the best practices of breeding with the best practices in business to create the 'business of plant breeding'.
- **Liaison with government officials.** Breeders need to work closely with key government officials and, where possible, sensitize them of the need for new varieties with traits that respond to market demands. By so doing, they aim to solicit official support for new demand-led varieties and advocate their use.
- **Compelling variety development cases.** Breeders need to understand clients' needs and create varieties that have benefits for all in the value chain and deliver an attractive return on investment. They also need to develop a broad and deep understanding of the range of costs required to develop demand-led varieties, to be able to create business investment cases that are persuasive to government officials and to both private and public investors, and to secure and retain support for a demand-led breeding programme/project.

## Key messages for plant breeders

- **The investment case.** Clarity is required on the rationale and justification for proceeding with a demand-led breeding programme. A compelling case for investment should be constructed. Investment cases are always assumption based. The quality of the case comes from detailed analysis of the benefits and the performance assumptions, including questioning their probability and understanding their sensitivity to factors such as farmer adoption, choice of the varieties available and changing variety development costs.
- **Clients.** It is critical to understand who the clients are, to define client segments and to determine whether there is a significant return on the proposed investment. If there is insufficient economic and social return, alternative crop improvement programmes should be considered and given higher priority.
- **Communication.** The creation of a compelling investment case that is persuasive to government officials, investors and stakeholders is critical to securing and retaining support for a demand-led breeding programme.

## Key messages for research and development (R&D) leaders, managers, government officials and investors

- **Return on investment.** Governments, R&D managers and investors need to appreciate that breeding programmes can provide a return on investment and need not only be seen as a budget cost. Managers need to encourage an investment decision-making culture rather than one of budget spending. This can be achieved by tracking the adoption of and benefits that have accrued from new varieties, in addition to simply monitoring the number of varieties developed by breeding teams that are registered for release.

- **Benefits estimation.** R&D management needs to provide expertise in agricultural economics and social sciences in order to work with the breeding team to construct a compelling investment case and quantify the likely benefits before a new breeding programme begins.

## Introduction

The objective of this chapter is to strengthen the ability of plant breeders to create compelling business cases for investments in demand-led plant breeding. The chapter discusses how to develop a clear vision of both the benefits and the costs of developing new crop varieties. It describes the elements necessary for plant breeders to be able to create a compelling investment proposal and present this to R&D management, government officials and financial investors. This investment case includes: identifying the benefits and intended beneficiaries of a proposed new breeding project/programme; understanding the principles of return on investment; and clarifying whether the proposed investment in demand-led breeding can be justified in terms of its economic, social and environmental benefits versus its costs.

Plant breeding as an endeavour is both a science and an art, and requires an in-depth understanding of the fundamental nature of plants and genetics, as well as of the most advanced molecular science and rapidly emerging new technologies that can speed up varietal development. In making the case for plant breeding, it is necessary both to mobilize public support for science as a contributor to progress in global food and agriculture and, specifically, to increase investment in plant science as a means of improving crop productivity, especially in tropical environments.

There are insufficient resources available for meeting the food security challenge in Africa. Public investments by African governments on science and technology, as well as investments by the emerging private sector, need to increase to meet these challenges (FARA, 2014). There is a heightened need for scientists to communicate and justify their work, and for this to be done using explanations that are persuasive to policy makers, politicians and private sector investors (BSPB, 2012). Thinking about the creation of new varieties within a framework of economic investment that generates a return for farmers, businesses and rural and urban communities in the countries of Africa is a way of shifting the agenda to the real landscape, in which plant breeding is seen to make a contribution to economic growth and livelihoods, rather than being seen only as a cost to the public purse.

National public breeding programmes currently receive the majority of their funding from their own governments. Public investments will continue to be required, in order to operate a long-term, sustainable breeding programme that has a critical mass of human and financial resources, which will enable it to deliver a pipeline of new demand-led varieties over time. This national support can be supplemented by international development assistance and by philanthropic foundations. Such external support tends to be provided over shorter duration and specific projects, sometimes focused on a single crop trait (e.g. drought tolerance or disease resistance). There are also opportunities to grow the role of the private sector in supporting demand-led plant breeding.

It is vital for professional breeders to develop the skills to win support for reliable, sustainable funding by critically analysing their programmes, demonstrating and communicating the merits of their investment cases, and then delivering varieties that achieve their promises for all of the beneficiaries involved. As such, the aim of the chapter is to act as a resource for education in this field. For this purpose, a box is included near the end of the chapter that summarizes the main educational objectives of the chapter and presents the key messages and questions involved. There is also a final box at the end of the chapter that summarizes its overall learning objectives.

## Investment Decisions

Demand-led new variety development puts clients and their value chain actors at the front and centre of the plant breeding programme. The first decision and entry point on starting a new breeding programme is deciding whether the demand by clients has sufficient scale and benefits to justify the investments and costs required. New varieties that are designed for clients can deliver broad-ranging benefits – from direct economic gains to social and environmental benefits. Benefits may also pertain to other stakeholders such as government officials, international development agencies and philanthropic foundations, as the new varieties will contribute towards achieving their official development policies and sustainable development goals.

### Benefits and beneficiaries

For any breeding programme, there must be target products to be developed – varieties – that are designed for specific clients, according to the established demand. The new varieties should have specific attributes that are unique or better than the attributes of existing varieties. Before developing these new varieties, it is important to understand the economics of so doing.

One aspect of the business case is to be able to quantify the benefits of developing a new demand-led variety and how that will influence the expected adoption rates. This will help decision makers to understand the potential value of the new variety. Comparing these likely benefits with those of current varieties, and understanding the external market and trends, are also essential ingredients in being able to prepare an investment case. Investment cases should strive to define benefits in both quantitative and qualitative terms, and take both into account.

The most compelling business cases for gaining support from government officials, investors and partners within the private sector are those that demonstrate economic returns combined with social and environmental benefits. Some key questions to consider here are:

- Who are the beneficiaries?
- What are the benefits that a new demand-led variety can deliver to the various beneficiaries?
- What are the individual benefits for each stakeholder in the value chain?
- How can one put a value on these benefits?
- Can the benefits be quantified and, if not, why not?
- Are there recognized valuation methods?

Often, the beneficiary who is overlooked when making the business case is the public and/or private financial investor in the plant breeding programme. New varieties can contribute to achieving both the policy goals and the economic growth targets set by governments, and those of private investors by gaining a return on their investment. In most countries of Africa, the investors in plant breeding are mainly the national government, international development agencies and/or private foundations. In countries with more mature seed markets, such as South Africa, the private sector is increasingly investing in plant breeding. This chapter focuses on preparing investment cases for both public and private funding sources.

A checklist of the benefits that a new demand-led variety might deliver is given in Table 7.1. This is not an exhaustive list of every aspect that may be important for value chain actors. It is intended as a guide to stimulate discussion on the different types of benefits that have value for investors, clients and other stakeholders, and that should form part of an investment case; it also aims to address some of the ways that these benefits can be quantified.

*Scale of the beneficiaries*

To be realistic about the potential value of new demand-led varieties, it is necessary to consider the potential number of farmers who will use each variety. The derived potential value must be compared with the cost of investment to develop the individual variety. When the investment in a breeding programme (e.g. by the private sector or an international research programme) is being considered, in the case where the investors may have interests in multi-country programmes, it is also necessary to consider potential users of the variety in other countries that share similar agroecologies, because this would expand the market potential for the new variety.

By investing in breeding in one country and supplying the variety to potential clients in other countries, the breeding programme (and its investors) is reducing the costs that would occur if the breeding activities were replicated in several countries. Investment in breeding a demand-led variety would, therefore, be more viable because of the increased number of potential users derived from cross-border trade; this applies even if the investment costs are higher than the value of the variety when only the potential users in the originating country are considered.

**Table 7.1.** Types of benefits gained from developing a new demand-led variety: examples of consequences and measurement units.

| Beneficiary | Specific benefit | Benefit consequence | Benefit type | Quantification units |
|---|---|---|---|---|
| Farmers | Greater yield | Farmer income<br>Shift from subsistence farming to crop surplus entering markets<br>Business growth<br>Cash income for children, education, health | Economic | US$ |
| | Earlier or later cropping than in the main season | Higher prices (as less supply out of season) | Economic | US$ |
| | Improved crop quality | Higher price, more customers | Economic | US$ |
| | Improved plant architecture | Easier harvesting<br>Time-saving | Economic | US$ |
| | Fewer inputs required (e.g. less fertilizer, insecticide) | Cost saving | Economic<br>Environmental sustainability | US$ |
| | Less labour | Cost saving | Economic | US$/person hours |
| | | Children have time to go to school | Social | Better literacy levels |
| Seed producers | Greater seed yield<br>Higher productivity per area grown | Farmer income<br>Unit costs less<br>More competitive price to distributors | Economic | US$/person hours |
| Transporters | Less damage in transit | Cost saving | Economic | US$ |
| Wholesalers | Improved shelf life | Cost saving | Economic | US$ |
| Food processing companies | Source from local farmers rather than imports | Cost saving<br>Reliable supply | Economic<br>Logistics | US$ |
| Food retailers/ Supermarkets | Good varieties and sourcing from local smallholders | Freshness and higher prices<br>Differentiation and fair trade brands | Economic | US$ |
| | Improved shelf life | Loss of wastage and costs | Economic | US$ |

*Continued*

**Table 7.1.** Continued.

| Beneficiary | Specific benefit | Benefit consequence | Benefit type | Quantification units |
|---|---|---|---|---|
| Consumers | Improved shelf life | Less time shopping<br>Less wastage and cost | Economic<br>Social | US$ |
| | Easier preparation | Time-saving | Economic<br>Social | US$ |
| | Shorter cooking time | Energy saving | Economic<br>Social | US$ |
| | Increased vitamin or protein content<br>Reduced mycotoxins<br>Improved health | Working ability<br>Reduced health costs | Economic<br>Social | US$ |
| Investors<br>Governments<br>International agencies<br>Private sector | Delivery of mandate/ policies<br>Support balance of payments (import versus export)<br>Economic development<br>Farmer livelihoods | Economic development<br>Continued funding for plant breeding projects and support for innovation and science | Economic<br>Social | US$ |

This proposition supports the business case for establishing regional programmes, and also underlies the efficiency of regional and international breeding programmes supported by the CGIAR international agricultural research centres and CGIAR research programmes (CRPs), and by regional biosciences facilities (such as Biosciences eastern and central Africa – BecA), subregional organizations in Africa (such as ASARECA, the Association for Strengthening Agricultural Research in Eastern and Central Africa; and CORAF/WECARD, Conseil Ouest et Centre Africain pour la Recherche et le Développement Agricoles/West and Central African Council for Agricultural Research and Development), as well as regional educational programmes for plant breeders (such as ACCI, the Africa Centre for Crop Improvement; and WACCI, the West Africa Centre for Crop Improvement).

#### *Identifying benefits: sources of information*

There are several sources of information that may help guide in identifying the types of benefits that can be used to construct a compelling investment case. These include the following.

GOVERNMENT DOCUMENTATION. National governments of OECD (Organisation for Economic Co-operation and Development) countries have their own evaluation systems for reviewing the performance of their science programmes, which often require impact studies.

For example, the Australian Grains Research and Development Corporation (GRDC) commissions independent assessment studies that provide comprehensive documents that review the economic impacts of its investment in plant breeding for individual crops. The GRDC is funded partially by farmer levies, which are matched with public funding. It uses these funds to support R&D for the Australian grains industry, including supporting breeding programmes for new variety development. The stakeholders expect value for their money, and hence a number of these documents are made publicly available at https://grdc.com.au/about/our-investment-process/impact-assessment (although some require permission for access). The GRDC evaluation approach is rigorous and systematic, and is a combination of quantitative and qualitative measures of economic, social and environmental benefits. It is a rich resource for breeders to consider using to help construct their own business cases in other crops and countries. A case study for consideration and discussion is the lentil breeding and export industry in Australia (GRDC, 2013).

Another useful resource document is a lobby paper submitted by the British Society of Plant Breeders to the UK Government on the importance of plant breeding, and its dependency on breeders' rights legislation to enable plant breeding investment by the private sector (BSPB, 2012).

INTERNATIONAL DEVELOPMENT AGENCIES AND ADVANCED RESEARCH INSTITUTIONS. Each agency or research institute has its own templates and requirements for how benefits should be communicated. Information can be found online on project proposal documentation and by seeking out *ex ante* impact studies that have been conducted by economists/social scientists.

PRIVATE SECTOR COMPANIES. Private sector plant breeding programmes are wholly dependent on delivering successful varieties that generate profits to satisfy shareholders, and they support investment in a variety development pipeline. Hence, substantial attention is given to identifying the economic benefits of such programmes to farmers, clients and the value chain as a whole. For example, Syngenta has made available a pro forma information tool that can be used to construct and evaluate investment cases based on their putative economic benefits. This tool is described in Appendix 7.1 at the end of the chapter.

## Investor expectations

When developing new demand-led varieties, the breeding team needs to be sure that it understands the investors' interests and expectations over the life of a demand-led breeding programme/project, as well as those of the clients and stakeholders. It is important that the investors should be part of the consultative process from the inception of the breeding programme, including during the process of developing the stage plan for the programme. This will help the programme to focus on the same areas as the investors' interests, while also addressing the needs of the clients who will grow and use the new varieties. The overriding consideration for publically funded breeding programmes in Africa is that they serve the needs of African smallholder farmers and other clients along the value chain.

## Cost Estimation and Management

Understanding the basics of project accounting and investment decision making is a core skill that can be beneficial to all public and private breeders and researchers. It is also an important skill that helps scientists to design and cost project proposals confidently and monitor these costs during project implementation.

Understanding the costs associated with developing new varieties will help in making key investment decisions in breeding programmes. This is required for demand-led programmes, because the inclusivity of clients and stakeholders may raise the need for additional studies that have a cost component. Partnering with value chain stakeholders may also provide additional resources, either 'in kind' or as direct funds for the programme.

Within the context of plant breeding, a holistic approach to project costs is possible because breeders start with an idea and have a tangible end product, i.e. a registered variety for use by farmers, for which costs can be calculated and assigned from the start to the end of the project, and the variety can have an estimated value.

There are many and varied costs associated with breeding programmes that need to be adequately considered. A full analysis is required for every cost that might be incurred, including staff time (in full time equivalents – FTEs) as well as operational funds (cash) and any capital or equipment costs. To act as a reference, a list of the costs that might be encountered during demand-led new variety development is given in Table 7.2.

### Cost calculation

The best way to calculate costs for a demand-led breeding project/programme is to use the stage plan (see Fig. 5.2, this volume) and the associated activity work plan for each new variety being developed, and then obtain the cost for every activity and assign an amount to each year in the timeline. This will require a dialogue with a range of budget holders, as some of the costs will not be from the breeder's own department. Some of the costs are also likely to be external to the research institute concerned. Consider all of the skills, activities and materials required, and the disciplines they are in. These costs could be measurable in people's time (FTEs) or in real cash costs, such as monies to be paid to a government department for variety registration trials and review by the variety release committee (Table 7.2). For the costs of people's time, calculations will require the use of a standard cost per person. This standard cost will be known or will need to be calculated, possibly by the finance department of the home institution. Usually, personnel costs are presented as a percentage of an FTE/year.

An aspect that requires a policy decision with the institute that is developing the variety is whether in calculating costs only the actual operational costs should be charged to a project budget, or whether the costs should include fully charged overhead costs in the estimate. Overheads in scientific laboratories with

**Table 7.2.** Reference table of breeding programme costs: examples of types of costs and their requirements for human resources (HR)-based costs (as full time equivalents, FTEs) and cash.

| Disciplinary area and cost item | HR input (FTE) | Cash |
|---|---|---|
| **Farmer and value chain market research** | | |
| Meetings and consultations with farmers and the value chain to define needs and priorities to create new variety designs and set breeding targets and goals | YES | YES |
| Specific market research studies | | |
| **Project governance and decision making** | | |
| Management meetings: to review project progress and make stage plan advancement decisions that include clients/ stakeholders in the decision making | YES | YES |
| Project management: to create the demand-led development plan, and monitor and evaluate progress | YES | YES |
| Investment case creation: discussions with economists, social scientists, management and budget holders to create the business case, which comprises project benefits and costs. Project proposal and plan creation, and liaison with investors | YES | YES |
| **Plant breeding** | | |
| Plant breeders | YES | YES[a] |
| Laboratory and greenhouse technicians | YES | NO |
| Molecular biology: sequencing, genotyping and other data analysis | YES | YES |
| **Experimental design and data management** | | |
| Bioinformatics advice and statistics packages | YES | YES |
| Computer access and power | YES | YES |
| **Germplasm evaluation** | | |
| Farm trial operations – labour (on-site, off-site) | YES | NO |
| Farmer participatory breeding trials | YES | YES |
| Agronomists | YES | NO |
| Plant protection | YES | NO |
| Soil scientists | YES | NO |
| Processing performance tests | YES | YES |
| Performance tests by food company or other value chain stakeholders | YES | YES |
| Consumer-based assays (including outsourcing), e.g. cooking and taste trials | YES | YES |
| **Procurement of supplies** | | |
| Procurement team | YES | NO |
| Purchase of inputs for trials (fertilizer, chemicals, seeds, water) | NO | YES |
| Import, export and transport of germplasm | YES | YES |
| **Seed production** | | |
| Scaling production for registration and demonstration | YES | YES |
| Seed certification | YES | YES |
| **Variety registration** | | |
| DUS, VCU, EOs trials[b] | NO | YES[c] |
| Review by the variety release committee (VRC) | NO | YES |

*Continued*

**Table 7.2.** Continued.

| Disciplinary area and cost item | HR input (FTE) | Cash |
|---|---|---|
| **Demand-creation – demonstrations, promotion and line selections at different times in the stage plan and at variety release** | | |
| Farmer groups | YES | YES |
| Extension service | YES | YES |
| Distributors | YES | YES |
| Value chain stakeholders (processors, supermarkets, etc.) | YES | YES |

[a]This includes travel costs for discussions and information seeking, which are additional to plant breeding costs per se.
[b]DUS, testing for distinctiveness, uniformity and stability; VCU, determining value for cultivation and use; EOs, variety testing trials by examination offices (EOs) performing on behalf of the community variety office.
[c]Registration trial and release costs can be substantial. For example, currently 2 years of mandatory trials for a new bean variety in Kenya costs US$3000 per variety. Additional years of trials, if required, cost US$1200 p.a.

expensive hi-tech equipment (that lose value over time, i.e. have a depreciation cost) can be as much as 30–60% and are material to an investment decision on a plant breeding programme. Most investors will accept some overhead costs, and the level and content of overheads in a programme usually need to be negotiated on a case-by-case basis.

In most countries, the national government's agricultural R&D programme has a budget (sometimes insufficient) for conducting crop improvement activities. The main decision to be made is one of the choice of which crops to select for the breeding programme. In this case, it is more appropriate to use non-overheaded costs. This is also the method to use in the Syngenta investment modelling tool, in which the non-overheaded costs are inserted into the model (Appendix 7.1). However, it is worth understanding that this approach underestimates the full cost of the breeding programme because these overheads are omitted. The full costs of running a breeding programme should include overheads that cover the essential institutional costs.

## Cost efficiency

With limited resources, it is appropriate to make rational decisions on their use, to maximize efficiency during the variety development process. Each breeding programme will be different. Some areas that are worth considering for cost efficiency are given below.

### *Internal capacity versus outsourcing expertise*

A key decision is the location and source of expertise and activities. A natural tendency is to prefer to do all activities within the home or partner institutes. This may not always be the best approach, especially for demand-led programmes that involve consumer- or processing-based assessments for which technical capacity or know-how is limited at the home institution and will take time and investment to acquire. Cost is also an important consideration. Outsourcing may be a better option for some assessments.

In recent years, because of the genomics revolution, the costs of sequencing and genotyping have reduced dramatically (Boettiger *et al.*, 2013). One area where breeding programmes have cut costs is by making efficient use of the available resources by outsourcing sequencing and genotyping services. One such example is the approach being used by the African Orphan Crops Consortium (AOOC; see http://africanorphancrops.org/). Beijing Genomics Institute (BGI) is doing the high-throughput primary sequencing of 100 lines each of 101 African crops in China. Resequencing and annotation work is being done at the World Agroforestry Centre (ICRAF) in Nairobi, together with the BecA-ILRI Hub at the International Livestock Research Institute (ILRI) in Nairobi, which is providing support and curation of the AOCC germplasm.

Another example of outsourcing involves considering the best approach for access to genetic markers. For some traits in the demand-led variety development programme, it may be better to access markers at high-throughput molecular laboratories. Tissue or DNA samples from a breeding programme can be sent to such central laboratories for genotyping at a reduced cost compared with running the same processes individually within the breeding programme. The results from genotyping are sent back to the respective breeding programme for analysis and selection of the lines or varieties that have the required genes. This becomes cost-effective as the breeding programme only pays for the services rather than investing in the laboratory infrastructure, chemicals and personnel to run it. Some breeders based at public institutions in Africa are accessing such outsourcing services by sending their samples for genotyping at K-Bio-Sciences in the UK through the Integrated Breeding Platform (IBP) of CGIAR's Generation Challenge Programme (GCP), with financial support from the Bill & Melinda Gates Foundation.

#### *Crop cycles per year*

Other cost efficiencies are also possible, especially where the number of crop cycles can be increased from one to two or even three cycles a year. This is possible in some cases, depending on the crop species, rainfall pattern and the availability of facilities such as irrigation or greenhouse space. Increasing the number of crop cycles a year cuts the time from generating crosses to the release of a variety. As time is associated with costs, reducing time will also reduce some costs. For example, there have been recent developments of barley and wheat that have used controlled environmental conditions to increase the number of cycles a year from one to seven. The approach has been developed by Hickey (2014) at the University of Queensland, Australia and is called 'speed breeding'.

Speed breeding has applicability for a range of crops but it is crop species dependent due to day-length requirements. Breeders who have access to controlled environment facilities are encouraged to explore this area of practical agronomy for use in their programmes.

#### *Demand-led approaches and client involvement*

There are various scenarios using the demand-led breeding approach that can have an impact on the costs of breeding programmes. The first scenario is that by involving various stakeholders with interests in the demand-led variety design and development, more interactions and processes are required. This might translate into increased costs, both directly and indirectly, due to the extra time required to accommodate the various processes demanded by different stakeholders. As such, the demand-led breeding approach raises the cost of producing a variety. However, the approach can produce varieties with reduced risks of failure and increased uptake by clients. These positive factors in the breeding programme will ultimately lead to reduced development costs per variety used by farmers.

The ideal scenario is where stakeholder engagement leads to clear design requirements, resulting in a more focused performance appraisal of lines, quicker decision making on the key traits, faster development and reduced costs. In some cases, it may be possible – where stakeholder interest in the value chain is particularly high – for a partnership to be developed where there is sharing of costs (e.g. the benefiting processor, or a food company, provides bioassays or performance testing arrangements, or other contributions with know-how, resources or finance).

*Fast-tracking development and risk management*

A key investment consideration is the speed of the project plan and the scale of the risk mitigation measures that are included. Generally, breeding and cost efficiency will be sought in most programmes owing to budget constraints and the need to register varieties as rapidly as possible to meet the needs of farmers, clients and investors. Fast-tracking may require additional resources to ensure success (e.g. the parallel testing of many lines for a longer time period, or doing additional work to reduce market uncertainty). Investment cases often show that in developed markets, reaching the market first (even by 1 or 2 years) with a highly differentiated trait can make a significant difference to the attractiveness of the investment proposal, and so additional upfront costs are acceptable in order to reduce the time to market for the new variety.

## Investment Decision Making

The decision to invest in developing a new variety is a strategic one for public and private sector institutions, because of the potential benefits for farmers and the value chain, the length of time it takes and the large costs involved. Decision making in public R&D organizations is often based on generating public goods, farmer benefits and breeding success (as measured by the numbers of varieties being registered rather than by adoption rates or breeding efficiencies and cost management (Ceccarelli, 2015).

The concept of generating a return from an investment in breeding has always been the driving mandate by the private sector, but is a relatively new way of thinking for public sector breeding programmes. A return on investment is displayed most simply as a multiple factor. For example, if the estimated economic benefit of a new variety is US$1 million and the total development costs are US$ 0.5 million, then the benefits have covered the cost of the investment and provided a return on investment of 100%.

The approach made by the private sector in developing a business case is driven by aspects such as strategic considerations on business growth and sustainability, the shape and size of variety portfolios, and a clear focus on the cost of financing for the variety development programme and the return on investment. Therefore, the private sector conducts rigorous analysis on the following aspects in making its investment decision:

- The performance and profile of a potential new variety and how it compares with current and alternative new varieties being developed in-house or by other organizations.
- The likely scale of sales and profitability, and the crop and market growth dynamics.
- All of the costs (both operational and capital) involved with the development and launch of the variety.
- Whether the investment needed in a new variety can help to sustain, and preferably to grow, the business?
- Whether all of the costs can be remunerated and whether there is an opportunity cost, i.e. is there an alternative variety design that would be a better investment?

It is common for seed companies in mature and competitive markets to replace their existing varieties with newer ones that have incremental benefits and so the investment case addresses not just the value of the new variety but also the degree of competition with the company's own brands (i.e. previously released varieties still in the market). For many crops and countries of Africa, there is still much scope for modern varieties to displace less well-performing landraces without competition with exiting varieties being a major consideration (Walker and Alwang, 2015). An exception is maize in South Africa and Kenya, where there is high adoption of modern (hybrid) maize varieties and much competition between seed companies in releasing new varieties.

## Discounted cash flow models

There is a range of methods that can be used to make financially based investment decisions on new product development (Artmann, 2009). Each has advantages and disadvantages. Agencies such as the Australian GRDC and private sector agribusinesses and seed businesses use discounted cash flow models as part of building their investment cases.

A key reason why discounted cash flow models are favoured is because this valuation method allows direct comparisons of projects with different dimensions, economic benefits and costs. For example, it allows a comparison of projects with large sales, high costs and many years to delivery with projects on a niche crop variety for a small number of farmers that can be registered quickly at a low cost (and every variation in between). The method is also used to help make go/no go decisions on individual varieties and as a portfolio tool to help rank and prioritize which projects progress within a breeding programme.

The underlying principles of the method are the same for all of the above iterations. In the case of plant breeding, some public investors also use discounted cash flow analysis, as well as metrics such as the internal rate of return for investment decisions. For example, USAID (the US Agency for International Development) uses net present value (NPV) analysis as part of its impact assessment approaches for investments in plant breeding programmes, and favours option-based models (Andoseh *et al.*, 2014).

### *Principles of discounted cash flow models*

- **Assumption-based models.** NPV discounted cash flow models are wholly assumption based and the outputs are only as good as the assumptions used. In many instances, their core benefit comes from enabling project teams to focus on the quality of the input parameters used to generate the forecasts, such as seed sales indicating farmer adoption, and R&D costs, as well as the collective decision to invest in the programme. The actual NPV, once it is positive, is less important than the rigour and quality of the discussions on key input parameters. Sensitivity analysis to different assumptions is usually a contribution to the decision-making process.
- **Decision choices.** These enable investment decisions on whether to invest money in an investment that generates a fixed and reliable interest rate of return (and/or the cost of the monies needed) or to invest in a development project such as creating and releasing a new variety.
- **Value of money.** The key premise is that money has a greater value now than in the future because of the many risks and uncertainties that may arise.
- **NPV.** Here, a single investment monetary value is generated called a net present value or NPV. The NPV takes into account all of the cash flows involved, their timing and the cost of capital/alternative investment interest options by using a discount rate. It uses the concept of the time value of money, but applies it to all of the cash flows of an investment decision (sales in and costs out) and translates them into today's money.
- **Discount rate.** The discount rate is used to encapsulate the interest rate of return that could be obtained by investing elsewhere and the investment risk factors. This discount rate will vary according to the policies of the home institution but, typically, for agricultural investments of this type it is between 10 and 25%.
- **Project prioritization.** Any NPV that is positive means that the investment creates value. This estimated value is higher than if the money was to be invested in an interest-bearing account (assuming that the assumptions prove to be correct). The absolute number has meaning, but is often best used as a tool for ranking the relative importance of different breeding project options. The higher the NPV, the more attractive the investment project.
- **Terminal value.** This is an estimate of the value of the variety or business at the end of the time period of the investment and takes future cash flows into account.

- **Total investment value.** This is the sum of the NPV and terminal value of the new variety.
- **Sunk costs.** A sunk cost is a cost that has already been incurred, and cannot be recovered. In the case of a breeding programme, the sunk costs describe the increase in NPV with time during the development timetable to the point at which the development budget is spent.

A key principle of investment and the use of discounted cash flow models is that they look forward. Once money has been spent, it is no longer relevant to an investment decision. The key question is always whether with the facts to hand you should spend more money – rather than reflecting on a previously made decision. It is human nature to think about the money that you have already spent, but this should not cloud future decision making. Cash flow models do not include monies that have already been spent; they address the time period for investment from now onwards.

### Other metrics

There are other metrics that can help in decision making on development investments. These include calculation of the payback period and the internal rate of return, which has already been mentioned.

#### *Payback period*

This is the time taken from the start of the investment to when the income stream from sales fully covers the development and other costs. The payback period can be very short if there is a large market, rapid penetration and a high demand for the new variety.

There are many examples of new varieties where the payback period is much longer (or is never reached), due to low adoption of the new variety by farmers as a result of poor design or lack of seed distribution and sales. Sometimes, the development costs are never fully recovered or justified. In all cases, there is a time lag between the requirement for financial resources to be spent and the money being recovered. Projects that have a shorter payback period are likely to be favoured, particularly when programmes have insecure longer term funding and investors are seeking quick results.

#### *Internal rate of return*

The internal rate of return (IRR) is the discount rate (%) that makes the NPV of all the cash flows equal to zero. Usually, the higher the IRR the more desirable it is to make the investment.

## Public and Private Investment in Plant Breeding

Currently, there is much attention paid by governments and research managers around the world on how to attract investment in plant breeding to meet the challenges of food security and poverty. This is especially the case in Africa given the lack of national investment in science and technology that has occurred in many countries over past decades (FARA, 2014).

Public–private partnerships are being encouraged and novel investment financing models being sought (FAO, 2013) for investment in plant breeding. There are other approaches that can be considered as well, all of which are dependent on a remuneration flow back to the breeding investors. In this respect, operational seed legislation that allows the ownership and use of germplasm through breeders' rights is a prerequisite.

Australia has an interesting investment model that encourages the private sector to invest in the breeding of commodity crops. It uses farmer levies to part finance pre-breeding and also has legislation on plant breeders' rights that enables specific variety royalties to flow back to private companies. For example, wheat is a major export crop and economic earner for Australia. However, by 2000, support for public breeding efforts on a major export crop that generated profits for private companies was winding down. The seed companies were not able to invest in breeding because they could not recoup their investments. In response, an innovative remuneration model was created by changing the seed legislation. The remuneration flow is managed as an 'end point royalty' (EPR) payment that has to be paid as a 'research' fee when the crop is sold, or as a 'departure tax' before exporters can leave international ports (GRDC, 2011; Jefferies, 2012). This has dramatically changed the shape of investment in wheat R&D in Australia and increased the commitment of the private sector to the development of new, market-demanded varieties with differentiated cooking quality.

The future of plant breeding and the emergence of the private sector in Africa are highly dependent on the decisions taken by governments on seed legislation, breeders' rights, intellectual property and regulation enforcement. Without sustainable remuneration systems, the private sector is unlikely to invest in crops other than those where they can create closed-loop systems in which growers contract to return harvested seed to the variety owner, or there are hybrid seeds that can be sold annually, or technology fees are possible.

Public plant breeding is currently essential to support productivity improvements for most staple food crops and all orphan crops in Africa. However, most orphan crops and indigenous African fruits and vegetables are not served by even the public sector breeding system. As the seed distribution sector grows in African countries, it is to be expected that a similar trend to that seen in OECD countries will occur and that private sector investment in plant breeding will follow.

Capacity building in Africa is taking place to train breeders to be able to serve both the public programmes and the private sector, and hence the need to understand the best practices in investment concepts in both sectors. Often, a starting point for the emergence of a local seed company start-up, or for international seed companies to open an office in an African country, is the recruitment of a breeder or experienced agronomist. This initial step enables access to the deep understanding of local breeders about variety performance, seed production, and local knowledge of farmers' practices, their seed supplies and needs.

**Box 7.1.** The business case for investment in new variety development.

**Purpose:** to enable breeders to prepare and communicate to clients, stakeholders and potential investors a persuasive business case for investing in a demand-led breeding programme/project to develop market-demanded varieties. The business case includes identification of the benefits and costs, and justification for the investment in comparison with other approaches.

**Educational objectives:**

- to enable breeders to create a compelling plant breeding investment proposal by understanding the principles of return on investment and whether the investment is justified in terms of the economic, social and environmental benefits versus the costs involved;
- to demonstrate a detailed understanding of the benefits and costs of the breeding programme/project using qualitative and quantitative measures; and
- to communicate and justify a plant breeding programme to management, government officials, and public and/or private financial investors.

**Key messages**

- It is critical to understand who the clients are, how many client group segments there are and whether there is a significant return on the proposed investment.
- If there is no significant return on investment, alternative approaches should be considered.
- Clarity is required on the rationale and justification for proceeding with the breeding programme and providing a compelling case for new investment.
- Investment cases are always assumption based, as they depend on the success of the project and the future environment. The quality of the case comes from analysing the benefits and performance assumptions, while questioning their probability and understanding their sensitivity to factors such as farmer adoption, choice of varieties available and changing variety costs.
- The creation of a compelling business case that is persuasive to government officials, senior management, and public and private investors is critical to securing and retaining support for a demand-led breeding programme.

**Key questions**

- What are the key benefits that the new varieties will provide?
- Who will benefit and how?
- What is the value of the seeds and the size of the seed market?
- What is the value of the new varieties to smallholder farmers?
- What are the maximum adoption rates estimated to be?
- How can plant breeding contribute to the improvement of farmer livelihoods and national economic growth?
- How much will the plant breeding programme/project cost?
- How can the use of existing resources be maximized?
- Are there cost efficiencies that can be made?
- Does the number of potential users of the new variety(s) justify the costs involved?
- Are there efficiencies possible in breeding new varieties that may both meet regional needs and serve farmers in more than one country, or by creating a breeding network to reduce costs or address the lack of breeding capacity in smaller countries?

*Continued*

**Box 7.1.** Continued.

- Who are the potential investors?
- What are their interests and expectations?
- To what extent should investor interests be taken into account in the breeding design?
- Will the demand-led approach increase, decrease or not affect the cost of the breeding programme?
- What are the alternative options to the products being developed? Are there other ways apart from breeding that are quicker, easier, cheaper or more effective for the value chain?
- What is the cost–benefit ratio of the proposed programme/project? What alternative breeding programme(s) could be considered to deliver the market-demanded varieties?
- What are the financial ways of valuing new varieties?

Demand-led approaches being adopted in breeding programmes can help to raise adoption levels providing that they show very close linkages with seed systems and how the varieties that are created will reach farmers. Breeders need to work closely with both government and emerging private sector seed companies towards this goal. This will build the confidence of governments and investors, and will create a higher likelihood of investment for the next generation of improved varieties.

## Learning Methods

Before this chapter concludes, a summary is provided in Box 7.2 of learning methods – together with assignment and assessment methods – for use with the main topics that have been covered in the chapter: Investment Decisions, Costs and Decision Making.

## Conclusion

Understanding the value and costs of investing in plant breeding programmes is essential. Detailed analysis is required on the merits of each case and of the strength and degree of certainty linked to each assumption made.

The best professional breeders and institutional R&D managers from both the public and private sectors will spend significant time on understanding the context, delivery and relative merits of their new varieties to farmers, the value chain and the costs required to run a successful breeding programme. Senior breeders and their managers will be communicating regularly with their national governments and with investors and making the case in favour of their own plant breeding programmes using the appropriate economic, social and environmental arguments.

**Box 7.2.** Learning methods, and assignment and assessment methods.

**Investment Decisions, Costs and Decision Making**

*Learning method*

- PowerPoint presentation on investment decisions, their benefits and the beneficiaries.
- Group discussion on analysing the full breadth of the benefits and how to quantify them, with examples from current breeding programmes on how to quantify benefits.
- A case study of the Australian lentil breeding programme, for a review of how to construct a benefits case. Available at: https://grdc.com.au/__data/assets/pdf_file/0027/158373/an-economic-analysis-of-grdc-investment-in-the-lentil-breeding-program.pdf.pdf
- Group discussion to prepare a template for all costs that could be incurred in a plant breeding programme using an example provided by a participant or resource person.
- Group discussion on a case study that identifies the key learning points; group decision on whether to proceed with a proposed new breeding project, based on the investment case made.
- Using the investment tool to learn and explore some of the principles of developing variety development investment cases (see Appendix 7.1).
- Discussion of reference papers on cost efficiencies.

*Assignment*

- Participants to prepare and present their own variety design and make a compelling business case for proceeding with the investment and the subsequent breeding programme. The audience/peer review group to wear 'different hats,' such as government officials, regulators, clients in the value chain, investors and farmers, to test the quality of the content and presentation of the case.

*Assessment*

- Investment case for each participant based on their own breeding programme and on a written and oral presentation of that assignment.

Plant breeders and R&D managers will understand various financial terms and know how to compare investment cases using both quantitative and qualitative measures, as well as using value creation and return on investment as the driving concepts. This will involve understanding the scale of return and, at a more advanced level, the principles and outputs from cash flow models and use of such models to make prioritization decisions on their R&D programmes and new variety portfolios.

## Resource Materials

The open-resource e-learning materials available for Chapter 7 include: (i) Appendix 2, an electronic copy (e-copy) of the Excel-based Breeding Investment Tool, 'Financial Business Case for Breeding Programs' developed by Syngenta, which is further described in Appendix 7.1 of this chapter; and (ii) a set of slides available for this chapter as part of Appendix 3 that summarize the chapter contents and provide further information. The e-learning material is available at http://www.cabi.org/openresources/93814 and also on a USB stick that is included with this volume.

## References

Andoseh, S., Bahn, R. and Gu, J. (2014) The case for a real options approach to ex-ante cost-benefit analyses of agricultural research projects. *Food Policy* 44, 218–226. Available at: http://pdf.usaid.gov/pdf_docs/pnaec758.pdf (accessed 16 May 2017).

Artmann, C. (2009) Chapter 2: Literature review. In: *The Value of Information Updating in New Product Development*. Springer, Berlin/Heidelberg, Germany, pp. 9–39. Available at: http://www.springer.com/cda/content/document/cda_downloaddocument/9783540938323-c2.pdf?SGWID=0-0-45-685409-p173876683 (accessed 10 May 2017).

Boettiger, S., Anthony, V., Booker K. and Starbuck C. (2013) *Public–Private Partnerships in Plant Genomics for Global Food Security*. Research paper commissioned from the GATD [Global Access to Technology for Development] Foundation by the International Development Research Centre (IDRC), Ottawa. Available at: https://www.researchgate.net/profile/Sara_Boettiger/publication/242329658_Public-Private_Partnerships_in_Plant_Genomics_for_Global_Food_Security/links/02e7e51cc813353378000000/Public-Private-Partnerships-in-Plant-Genomics-for-Global-Food-Security.pdf?origin=publication_detail (accessed 10 May 2017).

BSPB (2012) Science and Technology: Written evidence submitted by the British Society of Plant Breeders. Lobby paper to British Government on the need for investment in plant breeding. Available at: http://www.publications.parliament.uk/pa/cm201213/cmselect/cmsctech/348/348vw20.htm (accessed 10 May 2017).

Ceccarelli, S. (2015) Efficiency of plant breeding. *Crop Science* 55, 87–97 Available at: https://www.crops.org/publications/cs/pdfs/55/1/87 (accessed 10 May 2017).

FAO (2013) *Report of a Technical Consultation to Promote Public-Private Partnerships for Pre-breeding, 30–31 May 2013, Rome, Italy*. Food and Agriculture Organization of the United Nations, Rome. Available at: https://www.fao.org/3/a-at915e.pdf (html version accessed 16 May 2017).

FARA (2014) *Science Agenda for Agriculture in Africa (S3A): "Connecting Science" to Transform Agriculture in Africa*. Forum for Agricultural Research in Africa (FARA), Accra. Available at: http://faraafrica.org/wp-content/uploads/2015/04/English_Science_agenda_for_agr_in_Africa.pdf (accessed 2 May 2017).

GRDC (2011) End Point Royalties (EPR[s]). Fact Sheet, Grains Research and Development Corporation (GRDC), Canberra/Kingston, Australian Capital Territory. Available at: http://www.seednet.com.au/documents/End%20Point%20Royalties%20Fact%20Sheet.pdf (accessed 22 May 2017).

GRDC (2013) *An Economic Analysis of GRDC Investment in the Lentil Breeding Program*. GRDC Impact Assessment Report Series. Grains Research and Development Corporation (GRDC), Canberra, Australia. Available at: https://grdc.com.au/__data/assets/pdf_file/0027/158373/an-economic-analysis-of-grdc-investment-in-the-lentil-breeding-program.pdf.pdf (accessed 21 May 2017).

Hickey, L. (2014) The Speed Breeding journey: from garbage bins to Bill Gates. QAAFI Science Seminar, 28 October 2014, Queensland Alliance for Agriculture and Food Information. YouTube video available at: https://www.youtube.com/watch?v=5tsor4PuMmw (accessed 2 May 2017).

Jefferies, S. (2012) Cereal breeding and end point royalties in Australia. Paper presented at the FarmTech 2012 Conference, Edmonton, Canada, as reported by M. McArthur in

*The Western Producer* Feb. 9th, 2012. Available at: http://www.producer.com/2012/02/royalty-fee-based-on-production-attracts-breeders-%E2%80%A9/ (accessed 16 May 2017).
Walker, T. and Alwang, J. (eds) (2015) *Crop Improvement, Adoption and Impact of Improved Varieties in Food Crops in Sub-Saharan Africa*. CGIAR Independent Science and Partnership Council (ISPC) Secretariat, Food and Agriculture Organization of the United Nations (FAO), Rome, and CAB International, Wallingford, UK.

## Appendix 7.1: Investment Analysis Tool

An e-copy of this investment analysis tool, as Excel-based spreadsheets, is available as Appendix 2 of the open-resource e-learning material for this volume. The e-learning material is available at http://www.cabi.org/openresources/93814 and also on a USB stick that is included with this volume.

### Introduction

Syngenta AG has kindly provided an educational tool that can be used by experienced breeders to explore some of the key principles of investment analysis. The tool is a macro that can be used on any computer that can run Microsoft Excel software.

This planning tool provides an opportunity for breeders to think through their own projects and the economic justification for proceeding with them or not. The tool is to help participants explore the concepts of making the case for investment decisions, rather than being able to generate a *de novo* investment case for their own project – which would require adjustments to be made to the model and tailored inputs for each case.

The investment tool addresses only economic assumptions. Social, environmental and other sustainability benefits need to be approached in a complementary manner, using both qualitative and quantitative measurements. A case study for participants would be to seek a methodology to place a value on the community and social benefits of new plant varieties, and for these to be incorporated into a broader socio-economic model.

### How to use the investment tool

The cells in the two Excel spreadsheets are connected with mathematical equations. Users are required to insert their own figures, but only in the designated areas in the sheets and cells that are marked in yellow. The tool assumes there is an ongoing plant breeding programme and that a new variety is being launched in year 1 that will take market share from existing varieties. This reflects a typical situation within a private sector company with an ongoing business, and also many public breeding programmes that are regularly releasing new varieties of the same crop with incremental improvements.

## Inputs

The inputs required by a breeder for the two spreadsheets are summarized below, under Project rationale and Financial metrics.

### *Project rationale*

This development case rationale can either be completed as part of the Excel spreadsheet or in a Word document. Addressing these topics in detail will help a breeder to create and evaluate their own plant breeding case for investment. The information required is as follows:

- Project information and responsibilities
- Project rationale
  - Product concept
  - Business rationale and core advantages and benefits of the new variety
  - Target market (crop and geography)
  - Crop/market dynamics
  - Competitor activities and alternative varieties
  - Variety competition (internal competition within breeding programme)
  - Main risks
  - Risk mitigation
- Other assumptions
  - Gross profit
  - R&D costs

### *Financial metrics*

These are the main inputs that are needed from a breeding team to be included to run the model to generate the financial investment case and create the financial metrics:

- Discount rate: % (usually a set figure is used by an investor institution based on cost of capital and risks)
- Total seed market value: US$ '000
- Annual market growth rate: %
- Market share of new variety: %
- Annual loss of market share for existing business: % (the new variety replaces the current one)
- Gross profit of seed (sales less cost of goods/seed production costs): %
- Annual total breeding costs/year: US$ '000

## Outputs

The outputs from the calculations are shown in the summary page of the Excel spreadsheet. This displays a range of performance and investment metrics and graphs from the development case, as follows:

1. Net present value
2. Terminal value

**3.** Sensitivity analysis with two scenarios compared with the base case set of assumptions for sales and gross profit of ±20%.

- Sales: base case ±20%
- Gross profit: base case ±20%

**4.** Key performance indicators:

- Payback period
- Year of first sales (in this model fixed at year 1)
- Peak sales
- Gross profit
- Total development expense
- Capital expenditure (not included in this model)

**5.** Two graphs:

- Sales: showing incremental sales with new variety outselling existing varieties over time
- Development expenditure over time
- Cash flow (discounted at 16% and undiscounted)

Although capital expenditure (e.g. infrastructure costs as opposed to operational expenditures) is often required to develop and take new varieties to market, for the purposes of this example, it is assumed that public breeding institutions will not be making these capital expenditures for specific varieties. Therefore, capital expenditures have been left out of the model at this time. However, there is a place to include these numbers on the spreadsheet, if they are relevant to another example.

This summary page of data is used together with the project rationale as the basis of a decision on investment. It is a multifaceted decision, and is unlikely to rely on any single metric or piece of information. Clearly, the NPV must be positive, ideally with a short payback period, a good internal rate of return and little internal competition that would reduce the profits from existing varieties (i.e. the new variety helps to grow the overall crop business and market share), with rapid sales growth/farmer adoption, low development costs and seed production costs as feasible, so that the gross profit per unit of seed is high.

# Index